Report of Investigations 9659

Evaluation of Explosion-Resistant Seals, Stoppings, and Overcast for Ventilation Control in Underground Coal Mining

By Eric S. Weiss, Kenneth L. Cashdollar, and Michael J. Sapko

U.S. DEPARTMENT OF HEALTH AND HUMAN SERVICES
Public Health Service
Centers for Disease Control and Prevention
National Institute for Occupational Safety and Health
Pittsburgh Research Laboratory
Pittsburgh, PA

December 2002

ORDERING INFORMATION

Copies of National Institute for Occupational Safety and Health (NIOSH)
documents and information
about occupational safety and health are available from

NIOSH–Publications Dissemination
4676 Columbia Parkway
Cincinnati, OH 45226-1998

FAX:	513-533-8573
Telephone:	1-800-35-NIOSH
	(1-800-356-4674)
E-mail:	pubstaft@cdc.gov
Web site:	www.cdc.gov/niosh

This document is the public domain and may be freely copied or reprinted.

Disclaimer: Mention of any company or product does not constitute endorsement by NIOSH.

DHHS (NIOSH) Publication No. 2003-104

CONTENTS

Page

Abstract	1
Introduction	2
Experimental mine and test procedures	2
Mine explosion tests	2
Instrumentation	5
Air leakage determinations	6
Cementitious pumpable plug seals	7
Construction	7
Explosion and air leakage test results	10
Australian design seals, stoppings, and overcast	12
Construction of seals, stoppings, and overcast	12
Seals	13
Stoppings	17
Overcast	18
Explosion and air leakage test results	22
First explosion test (LLEM test 358)	22
Second explosion test (LLEM test 359)	24
Third explosion test (LLEM test 360)	24
Fourth, fifth, and sixth explosion tests (LLEM tests 361, 362, and 363)	26
Seventh explosion test (LLEM test 364)	27
Preloaded solid-concrete-block seal designs for friable rib conditions	29
Construction	29
Explosion and air leakage test results	31
Conclusions	32
Acknowledgments	33
References	34
Appendix A.—Summary tables of air leakage measurements	35
Appendix B.—Summary tables of static pressure data for LLEM explosion tests	39
Appendix C.—Summary table of flame arrival data for LLEM explosion tests	45
Appendix D.—Summary tables of LVDT displacement data for LLEM explosion tests	46

ILLUSTRATIONS

1. Plan view of the Lake Lynn Experimental Mine	3
2. Seal test area in the LLEM	3
3. LVDT attached to a seal	5
4. Support posts and instrumentation on the back side of a seal	5
5. Pressurized entry for leakage determination rates across the seals	6
6. Brattice in place for seal leakage test	7
7. Construction of the wood and brattice cloth form walls used to contain the pumpable cementitious grout slurry	8
8. Slurry injection using the three injection ports located near the mine roof	8
9. Completed ribfill seal in crosscut 3	10
10. Horizontal cracks evident near the mine roof on the ribfill seal in crosscut 3 after test 354	11
11. Schematic of vinyl bladder with internal baffles used for construction of the seal and overcast designs	14
12. Spreader bar anchored to the mine roof used to support the seal bladder system	14
13. Shotcreting of the spreader bar and hook assembly	14
14. Inflated vinyl bladder assembly showing the injection port for the piers	14
15. Framework for construction of seal in crosscut 2	15
16. Construction of seal in crosscut 2 showing vinyl bladder in place, but not yet filled	15
17. Completed seal in crosscut 2	15
18. Construction of seal in high roof section of crosscut 3, showing the vinyl bladder being installed	16
19. Construction of seal in high roof section of crosscut 3, showing the grout injection hose attached to the bladder	16

CONTENTS—continued

Page

20. Construction of seal in crosscut 4 showing the vinyl tubes before filling 16
21. Construction of second seal in high roof section of crosscut 3 16
22. Construction of seal in crosscut 4 showing details of the tops of the vinyl tubes and light meshing overlay 17
23. Construction of water stopping in crosscut 3, with the individual tubes suspended from the roof-mounted spreader bar 17
24. Details of the Velcro and plastic clip fastening system for the water stopping in crosscut 3 18
25. Schematic drawing of overcast at the intersection of B-drift and crosscut 3 18
26. Construction of side wall of overcast at the intersection of B-drift and crosscut 3 19
27. Construction of overcast at the intersection of B-drift and crosscut 3: installation of deck on top of side wall .. 19
28. Construction of overcast at the intersection of B-drift and crosscut 3: installation of skirt around edge of deck 20
29. Top view of overcast deck showing reinforcing bars 20
30. View underneath the overcast deck showing temporary supports while the deck cement cured 20
31. Construction of wing wall along edge of overcast deck: side view of vinyl bladder for wing wall being installed ... 21
32. Construction of wing wall along edge of overcast deck: end view of wing wall above deck 21
33. Completed overcast viewed from under the deck 21
34. Side wall of overcast, as viewed from B-drift outby, showing instrumentation boxes and support frames 21
35. Instrumentation on top of overcast deck: three LVDTs suspended above deck 21
36. Closeup of LVDT suspended from roof and attached to deck 21
37. Completed water stopping in crosscut 3 22
38. Release of individual water tubes of stopping during air leakage test 22
39. Air-inflated vinyl bladder of quickseal in crosscut 4 22
40. Condition of water stopping in crosscut 3 after test 358 23
41. Completed seal in the high roof section of crosscut 3 23
42. Pressure traces as a function of distance from the closed end (face) in C-drift for test 359 24
43. Seal in crosscut 2 after test 359 24
44. Seal in crosscut 3 after test 359 25
45. Seal in crosscut 2 after test 360 25
46. Pressure and LVDT traces for seal 2 during test 360 26
47. Remains of crosscut 3 seal after test 360 26
48. Pressure and LVDT traces for seal 3 during LLEM test 360 26
49. Pressure and LVDT traces for overcast during LLEM test 363 27
50. Completed new (second) seal in high roof section of crosscut 3 28
51. Completed new seal in crosscut 4 28
52. Remains of crosscut 4 seal after test 364 28
53. Unfilled Packsetter bags at the seal interface with the mine roof and ribs 30
54. Filled and pressurized Packsetter bags at the outby roof and rib seal interface showing full-size bags and one half-size bag on the left 30
55. Placement of the Packsetter bag at the mine rib and floor interface with the bottom course of the tongue-and-groove solid-concrete-block seal 30
56. Hand-powered pump for filling the Packsetter bags 30
57. Completed mortared seal with the Packsetter bags and floor hitching in crosscut 2 31
58. Completed mortared seal with the Packsetter bags in crosscut 3 31
59. Mortared seal with floor hitching and Packsetter bags in crosscut 2 after test 366 32
60. Mortared seal with Packsetter bags in crosscut 3 after test 366 32
61. Remains of the dry-stacked seal with the Packsetter bags in crosscut 4 after test 366 32

TABLES

1. Lake Lynn Experimental Mine explosion tests 4
2. Guidelines for air leakage through a seal 7
3. Construction schedule at the Lake Lynn Experimental Mine 9
4. Seals and stoppings size data 9
5. Evaluations of the seal, stopping, and overcast designs 11

CONTENTS—continued

Page

A-1. Air leakage measurements before the first explosion test (No. 354) of the HeiTech program 35
A-2. Air leakage measurements after the first explosion test (No. 354) of the HeiTech program 35
A-3. Air leakage measurements after sealant was reapplied and before the second explosion test (No. 355) of the HeiTech program 35
A-4. Air leakage measurements after the second explosion test (No. 355) of the HeiTech program 35
A-5. Air leakage measurements before the first explosion test (No. 358) of the Barclay Mowlem program 36
A-6. Air leakage measurements between the first (No. 358) and second (No. 359) explosion tests of the Barclay Mowlem program 36
A-7. Air leakage measurements between the second (No. 359) and third (No. 360) explosion tests of the Barclay Mowlem program 36
A-8. Air leakage measurements between the third (No. 360) and fourth (No. 361) explosion tests of the Barclay Mowlem program 36
A-9. Air leakage measurements between the fourth (No. 361) and fifth (No. 362) explosion tests of the Barclay Mowlem program 36
A-10. Air leakage measurements before the seventh explosion test (No. 364) of the Barclay Mowlem program 37
A-11. Air leakage measurements after the seventh explosion test (No. 364) of the Barclay Mowlem program 37
A-12. Air leakage measurements before the first explosion test (No. 365) of the Packsetter seal program with the solid concrete block 37
A-13. Second air leakage measurements before the first explosion test (No. 365) of the Packsetter seal program with the solid concrete block 37
A-14. Air leakage measurements between the first (No. 365) and second (No. 366) explosion tests of the Packsetter seal program with the solid concrete block 38
A-15. Air leakage measurements after the second explosion test (No. 366) of the Packsetter seal program with the solid concrete block 38
B-1. HeiTech pumpable cementitious seals evaluation in the Lake Lynn Experimental Mine: pressure data, test 354   39
B-2. HeiTech pumpable cementitious seals evaluation in the Lake Lynn Experimental Mine: pressure data, test 355   39
B-3. Barclay Mowlem seal and stoppings evaluation in the Lake Lynn Experimental Mine: pressure data, test 358 40
B-4. Barclay Mowlem seals evaluation in the Lake Lynn Experimental Mine: pressure data, test 359 40
B-5. Barclay Mowlem seals evaluation in the Lake Lynn Experimental Mine: pressure data, test 360 41
B-6. Barclay Mowlem seals and overcast evaluation in the Lake Lynn Experimental Mine: pressure data, test 361 41
B-7. Barclay Mowlem seals and overcast evaluation in the Lake Lynn Experimental Mine: pressure data, test 362 42
B-8. Barclay Mowlem seals and overcast evaluation in the Lake Lynn Experimental Mine: pressure data, test 363 42
B-9. Barclay Mowlem seals and overcast evaluation in the Lake Lynn Experimental Mine: pressure data, test 364 43
B-10. Packsetter solid-concrete-block seals evaluation in the Lake Lynn Experimental Mine: pressure data, test 365   43
B-11. Packsetter solid-concrete-block seals evaluation in the Lake Lynn Experimental Mine: pressure data, test 366   44
C-1. HeiTech and Packsetter seals evaluation in the Lake Lynn Experimental Mine: flame arrival time data 45
C-2. Barclay Mowlem seals, stoppings, and overcast evaluation in the Lake Lynn Experimental Mine: flame arrival time data 45
D-1. Barclay Mowlem seals evaluation in the Lake Lynn Experimental Mine: LVDT data, test 358 46
D-2. Barclay Mowlem seals evaluation in the Lake Lynn Experimental Mine: LVDT data, test 359 46
D-3. Barclay Mowlem seals evaluation in the Lake Lynn Experimental Mine: LVDT data, test 360 46
D-4. Overcast LVDT data, test 361 47
D-5. Overcast LVDT data, test 362 47
D-6. Overcast LVDT data, test 363 47
D-7. Seal LVDT data, test 363 47
D-8. Overcast LVDT data, test 364 48
D-9. Seal LVDT data, test 364 48

UNIT OF MEASURE ABBREVIATIONS USED IN THIS REPORT

cfm	cubic foot per minute	lb/ft^3	pound per cubic foot
cm	centimeter	m	meter
cm^2	square centimeter	m^2	square meter
ft	foot	m^3	cubic meter
ft^3	cubic foot	m^3/min	cubic meter per minute
g/m^3	gram per cubic meter	min	minute
hr	hour	mm	millimeter
in	inch	MPa	megapascal
in H_2O	inch of water	ms	millisecond
kg	kilogram	psi	pound (force) per square inch, gauge
kg/m^3	kilogram per cubic meter	psia	pound per square inch, absolute
km	kilometer	psi-s	pound per square inch - second
kN-s	kilonewton second	s	second
kPa	kilopascal	t	metric ton
kPa-s	kilopascal second	V dc	volt, direct current
L	liter	°C	degree Celsius
lb	pound	°F	degree Fahrenheit

EVALUATION OF EXPLOSION-RESISTANT SEALS, STOPPINGS, AND OVERCAST FOR VENTILATION CONTROL IN UNDERGROUND COAL MINING

By Eric S. Weiss,[1] Kenneth L. Cashdollar,[2] and Michael J. Sapko[3]

ABSTRACT

A fundamental safety research area for the National Institute for Occupational Safety and Health (NIOSH) is to eliminate the occurrence of coal mine explosions or to mitigate their effects. One approach is to develop and evaluate new and innovative seal designs that provide increased explosion protection for mining personnel. The NIOSH Pittsburgh Research Laboratory (PRL) cooperated with HeiTech Corp. of Virginia, Barclay Mowlem Construction Ltd. of Queensland, Australia, and the Mine Safety and Health Administration (MSHA) in three separate research programs to evaluate the strength characteristics and air leakage resistance of numerous innovative seal designs and ventilation control structures for use in underground coal mines. For each phase of the program, various full-scale seals, stoppings, and an overcast design were constructed in PRL's Lake Lynn Experimental Mine near Fairchance, Fayette County, PA. The seals and stoppings were built in crosscuts and were subjected to explosions to evaluate their strengths.

Four pumpable cementitious seal designs ranging in thickness from 610 to 915 mm (24 to 36 in) were evaluated in the first cost-reimbursable research program with HeiTech Corp. A simple wooden framework with brattice liner was used as a form to contain the cementitious slurry during the curing period. As the seal designs decreased in thickness, higher compressive strength cementitious grout was used. All four seals withstood an explosion pressure pulse of at least 138 kPa (20 psi) while maintaining acceptable air leakage resistance.

In the second cost-reimbursable research program with Barclay Mowlem Construction Ltd. of Australia, several innovative seal and stopping designs and an overcast design were evaluated. Each of the seal designs and the overcast side and wing walls used one or more air-inflated vinyl bladder assemblies anchored to the mine roof and hitched into the ribs and floor. The air within these bladders was displaced with a high-strength cementitious grout. The overcast deck consisted of a 200-mm (7.8-in) thick reinforced cementitious slab. This was the first time that an overcast structure had been explosion tested under full-scale conditions. All of the seals, stoppings, and overcast design passed the air leakage tests before being subjected to a series of explosions with static pressure pulses ranging from 14 to 475 kPa (2 to 69 psi). Instrumentation measured seal and overcast wall displacement as a function of time. The 450-mm (17.7-in) thick seal in the 2.8-m (9-ft) high third crosscut withstood an explosion pressure of 170 kPa (25 psi), but failed during a later test, which generated a peak static pressure of 475 kPa (69 psi) at the seal location. A similar 450-mm-thick seal in the 2.1-m (7-ft) high second crosscut withstood three explosion tests, which generated peak static pressures of 195, 205, and 370 kPa (28, 30, and 54 psi) at the seal location. Next, the overcast design withstood four explosions, which generated static overpressures ranging from 16 to 47 kPa (2.3 to 6.8 psi).

A third program at the request of MSHA evaluated the effectiveness of using pressurized grout bags (Packsetter bags) along the mine roof and ribs in lieu of floor and rib hitching for a standard-type solid-concrete-block seal. This program was initiated to address an unusual geological mining condition encountered when building seals in entries where required rib hitching is not a viable option due to soft friable coal. Results showed that the use of these quick-setting grout-filled Packsetter bags pressurized internally to 300 kPa (44 psi) not only provides a seal that can withstand a 138-kPa static pressure explosion, but also provides a sealing option where rib hitching is not possible.

---

[1]Supervisory mining engineer.
[2]Research physicist.
[3]Senior research physical scientist.
Pittsburgh Research Laboratory, National Institute for Occupational Safety and Health, Pittsburgh, PA.

INTRODUCTION

During the course of underground coal mining it is sometimes necessary to install seals to isolate abandoned or worked-out areas of a mine. This eliminates the need to ventilate those areas. Seals are also used to isolate fire zones or areas susceptible to spontaneous combustion. To effectively isolate areas within a mine, a seal must—

• Minimize leakage between the sealed area and the active mine workings so as to prevent toxic and/or flammable gases from entering the active workings;
• Be capable of preventing an explosion initiated on one side from propagating to the other side; and
• Continue its intended function for 1 hr when subjected to fire conditions.

30 CFR[1] 75.335 [1997] requires a seal to "withstand a static horizontal pressure of 20 pounds per square inch (138 kPa)." Previous research by the U.S. Bureau of Mines (USBM) [Mitchell 1971] indicated that it would be unlikely for overpressures >138 kPa to occur very far from the explosion origin provided that the area on either side of the seal contained sufficient incombustible and minimal coal dust accumulations. This regulation formed the basis for previous evaluations of explosion-resistant seals at the Lake Lynn Experimental Mine (LLEM) [Stephan 1990a,b; Greninger et al. 1991; Weiss et al. 1993a,b,c; 1996; 1997; 1999].

The Pittsburgh Research Laboratory (PRL) and the Mine Safety and Health Administration (MSHA) have been jointly evaluating the capability of various seal construction materials and designs to meet or exceed the requirements of the CFR. This work is in support of PRL's Disaster Prevention and Response Research Program to improve safety for underground mine workers. These have been the first full-scale research programs to evaluate seal designs in entry geometries similar to those of current U.S. underground coal mines. Past seal research program had addressed, through explosion testing at the LLEM, the integrity of solid-concrete-block seals [Stephan 1990a; Greninger et al. 1991], low-density cementitious block seals [Stephan 1990b; Weiss et al. 1993c], cementitious foam seals [Stephan 1990a; Greninger et al. 1991; Weiss et al. 1993c], wood-block seals [Weiss et al. 1993c], cellular concrete seals [Weiss et al. 1996], and polymer seals [Weiss et al. 1996]. The overall objective of this research is to determine whether seals built from various materials and designs can withstand a 138-kPa explosion pressure pulse without losing their structural integrity. The seal must not only be physically strong, but also minimize air leakage. A safety benefit will also result from these evaluations in that many of these new seal designs require less materials handling and fewer worker-hours to install than the standard-type solid-concrete-block seal.

This report discusses the construction techniques, testing methods, and explosion test data collected for the pumpable cementitious seals; the seal, stopping, and overcast designs for the Australian program; and the Packsetter preloaded solid-concrete-block seal.

EXPERIMENTAL MINE AND TEST PROCEDURES

MINE EXPLOSION TESTS

All of the explosion and air leakage determination tests were done at the LLEM [Mattes et al. 1983; Triebsch and Sapko 1990]. Lake Lynn Laboratory is located about 80 km southeast of Pittsburgh, near Fairchance, Fayette County, PA. It is one of the world's foremost mining laboratories for conducting large-scale health and safety research. The LLEM is unique in that it can simulate current U.S. coal mine geometries for a variety of mining scenarios, including multiple-entry room-and-pillar mining and longwall mining.

Figure 1 shows a plan view of the LLEM. The underground entries consist of about 7,620 m of old workings developed in the mid-1960s for the commercial extraction of limestone and 2,370 m of new entries developed by the USBM in 1980-81 for research [Mattes et al. 1983]. These more recent entries are depicted in figure 1 as drifts A through D, each of which is ~520 m long and closed at the inby end, and drift E, which is 152 m long and connects drifts C and D. The drifts and crosscuts range from 5.5 to 6.0 m wide and are about 2 m high. The LLEM was designed to withstand explosion overpressures up to ~700 kPa (100 psi). During 1982-2001, a total of 406 consecutively numbered explosion tests were conducted at the LLEM.

Figure 2 shows a closeup view of the seal test area in the multiple-entry section of the LLEM. All of the seals and stoppings were built in the crosscuts between B- and C-drifts. The roof in one section of crosscut 3 had been enlarged previously to ~2.8 m high to more closely represent those of typical Australian and some U.S. underground coal mines [Weiss et al. 1999]. The roof in the intersection of B-drift and crosscut 3 was enlarged during this series of tests to accommodate the overcast design. Details on the designs for the seals, stoppings, and overcast are found in the "Construction" section for each of the three programs found later in this report.

Before each explosion test, a 60-t hydraulically operated, track-mounted, concrete and steel bulkhead was positioned across E-drift to contain the explosion pressures in C-drift

---
[1]*Code of Federal Regulations.* See CFR in references.

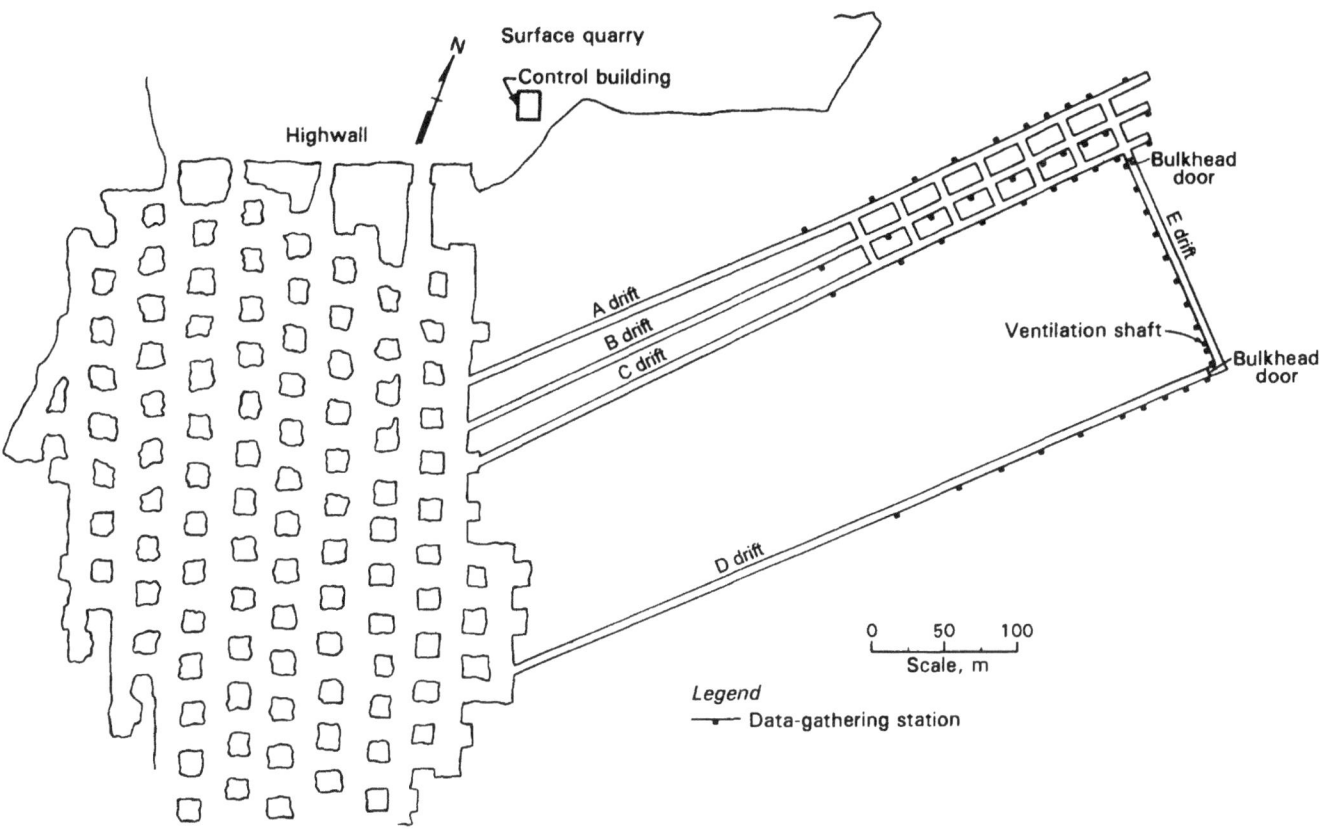

Figure 1.—Plan view of the Lake Lynn Experimental Mine.

(figure 2). For a typical evaluation test on a seal design for use in a U.S. coal mine, 18.7 m^3 (661 ft^3) of natural gas (~97% CH$_4$) was injected into the closed end of C-drift. An electric fan with an explosion-proof motor housing was used to mix the natural gas with the air in the ignition zone. A plastic diaphragm was used to contain the natural gas and air mixture within the first 14.3 m of the entry, resulting in a ~210-m^3 gas ignition zone. Sample lines within the ignition zone were used to continuously monitor the gas concentrations using an infrared analyzer. In addition, samples were collected in evacuated test tubes and sent to the PRL analytical laboratory for more accurate analyses using gas chromatography (GC). The GC analyses verified the infrared analyzer readings of ~9% of methane in air. Three electrically activated matches, in a triple-point configuration equally spaced across the face (closed end) of the entry, were used to ignite the flammable natural gas and air mixture. Barrels filled with water were located in the gas ignition zone to act as turbulence generators to achieve the projected ~140-kPa (20-psi) pressure pulse. The pressure pulse generated by the ignition of this methane-air zone generally resulted in static pressures ranging from ~150 kPa at crosscut 1 to ~115 kPa at the most outby location (in some instances as far outby as crosscut 5, or 150 m from the ignition source). Explosion studies have shown that the explosion pressure pulse decays less rapidly with distance in the larger LLEM entries (~13-m^2 cross section) than in smaller entries such as in PRL's Bruceton Experimental Mine (~5-m^2 cross section), presumably because of the smaller surface-to-volume ratio at the LLEM [Sapko et al. 1987].

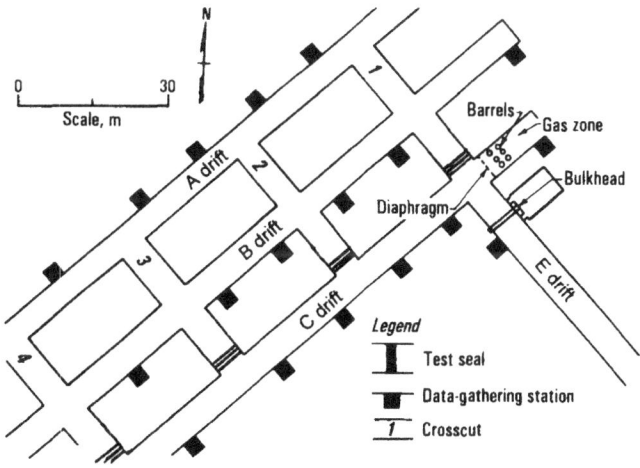

Figure 2.—Seal test area in the LLEM.

Summary data for the 11 explosion tests from the programs discussed in this report are found in table 1. In the table, the tests are identified chronologically within each of the three

Table 1.—Lake Lynn Experimental Mine explosion tests

Test No.	Date	Average maximum pressure[1]		Average flame speed[2]	Fuel type
		psi	kPa	m/s	
HEITECH PROGRAM					
354	Nov. 6, 1997	24	165	NA	18.7-m^3 CH$_4$ + 14.5-kg coal.
355	Nov. 20, 1997	23	160	~190 (26-71 m)	18.7-m^3 CH$_4$ + 14.5-kg coal.
BARCLAY MOWLEM PROGRAM					
358	Feb. 11, 1998	3.2	22	—	8.2-m^3 CH$_4$.
359	Feb. 27, 1998	27	185	~225 (26-93 m)	18.7-m^3 CH$_4$ + 14.5-kg coal.
360	Mar. 3, 1998	63	435	~385 (26-93 m)	18.7-m^3 CH$_4$ + 120-kg coal.
361	Mar. 26, 1998	2.7 [3]2.3	19 [3]16	—	8.2-m^3 CH$_4$.
362	Mar. 31, 1998	5.1 [3]4.3	35 [3]30	—	9.0-m^3 CH$_4$.
363	Apr. 1, 1998	7.7 [3]6.8	53 [3]47	~160 (26-41 m)	9.6-m^3 CH$_4$.
364	Apr. 3, 1998	23 [3]6.0	160 [3]41	~340 (26-56 m)	18.7-m^3 CH$_4$ + 7.3-kg coal.
PACKSETTER SEALS WITH SOLID-CONCRETE-BLOCK					
365	June 22, 1998	20	140	~380 (26-41 m)	18.7-m^3 CH$_4$.
366	June 25, 1998	23	160	~190 (26-71 m)	18.7-m^3 CH$_4$ + 14.5-kg coal.

[1]Average static pressures calculated in C-drift from 26 m to transducer just beyond last seal.
[2]Average flame speed calculated over distances (m) in C-drift, as noted in parentheses.
[3]Static pressure in crosscut 3, leading to overcast.

programs and also by the LLEM test number. Most of the tests (354, 355, 359, 364, 365, and 366) were set up in a similar manner to that described in the above paragraph, with a ~210-m^3 gas ignition zone. To ensure that all of the seal designs would see at least a 140-kPa explosion pressure pulse, a small amount of coal dust was used for several of these tests in addition to the gas ignition zone (see last column of table 1). The coal dust was loaded onto shelves suspended from the mine roof on 3-m increments starting 13 m from the closed end (near the end of the gas ignition zone). For tests 354 and 355 of the HeiTech program, test 359 of the Barclay Mowlem program, and test 366 of the Packsetter program, 14.5 kg of coal dust was loaded equally onto four shelves from 13 to 23 m from the closed end. The nominal coal dust concentration of this ~12-m-long dusted zone was ~100 g/m^3. When ignited, this coal dust increased the average explosion overpressure from ~140 kPa for the pure gas zone (test 365 plus some earlier tests) to ~166 kPa (24 psi) for tests with the hybrid gas-dust ignition zone. During the seventh Barclay test (LLEM test 364), only 7.3 kg of coal dust was loaded onto two shelves at 13 and 17 m from the closed end. When ignited, this ignition zone resulted in a slightly lower explosion overpressure. The average explosion pressures in table 1 were calculated from the pressure transducers in C-drift from 26 m to just beyond the last seal tested. For explosions with pressures >140 kPa (20 psi), the pressures in table 1 were rounded to the nearest 5 kPa (1 psi). The average flame speeds listed in table 1 were calculated from the flame arrival times listed in appendix C.

To achieve an explosion pressure pulse significantly in excess of 140-166 kPa, a larger quantity of coal dust was placed on shelves for a longer distance outby the gas ignition zone in C-drift. During the third explosion test (LLEM test 360) of the Barclay Mowlem program, a 64-m-long zone of coal dust (13 to 78 m from the closed end) was used in addition to the gas ignition zone. The pulverized coal dust (Pittsburgh Seam bituminous) was loaded onto the shelves to provide a nominal coal dust concentration of 150 g/m^3; this assumed a uniform dispersion of the coal dust over the entire cross section of the mine entry. A total of 120 kg of coal dust was used during this third seal evaluation test. This gas and coal dust mixture produced an explosion with an average overpressure of 435 kPa (63 psi).

To achieve the low explosion pressures (<70 kPa) necessary to evaluate the stopping and overcast designs during the first, fourth, fifth, and sixth tests (LLEM tests 358, 361-363) of the Barclay Mowlem program, the length of the gas ignition zone was reduced from 14.3 to only 8.2 m from the closed end of C-drift, giving an ignition volume of ~115 m^3. During tests 358 and 361, 8.2 m^3 (290 ft^3) of natural gas was injected within the gas zone, giving a methane concentration of ~7%. When ignited, the resulting gas explosions produced average overpressures in C-drift of ~20 kPa (3 psi). The small decrease in the average overpressure from test 358 to test 361 was mainly due to the number and types of ventilation structures being evaluated at the time. During test 362, 9.0 m^3 (319 ft^3) of natural gas was used and resulted in an average explosion overpressure in C-drift of about 35 kPa (5.1 psi). During test 363, 9.6 m^3 (340 ft^3) of natural gas was used and resulted in an average explosion overpressure in C-drift of about 53 kPa (7.7 psi). The pressures in the overcast area were somewhat lower, as noted in table 1.

INSTRUMENTATION

Each drift has 10 environmentally controlled data-gathering stations (shown in figures 1 and 2) inset in the rib wall. Each data-gathering station houses a strain gauge pressure transducer and an optical sensor to detect the flame arrival. The pressure transducer is perpendicular to the entry length and therefore measures the static pressure generated by the explosion. The pressure transducers were from Dynisco, Viatran, or Genisco. They were rated at 0-100 psia, with 0-5 V output, infinite resolution, and response time <1 ms. The flame sensors used Texas Instruments Type LS400 silicon phototransistors, with a response time of the order of microseconds. These phototransistors were positioned back from the front window of the flame sensors in order to limit the field of view.

Although the pressure transducers measured absolute pressure, the local atmospheric baseline pressure was subtracted from the outputted data traces so that they were gauge pressure values. For some of the explosion tests, the static pressure pulses exerted on each seal were measured by interpolation of the data from the two nearest C-drift pressure transducers, one inby and the other outby the crosscut position. However, in the later Barclay tests (LLEM tests 361-364), an additional transducer was installed in the rib at 75 m (246 ft) from the face, just opposite the seal in crosscut 3. Additional pressure transducers were installed on the C-drift (explosion) side of the seals and/or stoppings in crosscuts 1 through 4. These transducers were suspended about 0.45 m from the mine roof and were located about 0.3 m in front of each stopping/seal. They were positioned perpendicular to the seals. The pressure data recorded by these transducers would therefore be approximately the total pressure (combination of static and dynamic pressures) generated on the stoppings/seals during each of the explosion tests. The reason that the pressures may not have been quite equal to the true total pressures was that the housings were not designed in an ideal manner to measure the dynamic part of the total pressure. The "total" pressure data from these transducers located in front of the stoppings/seals were higher than the interpolated static pressure data.

An additional type of sensor was used during the Barclay Mowlem and Packsetter seal evaluation programs: linear variable differential transducers (LVDTs) to measure movement of the seals. The LVDT is shown attached to the back (B-drift side) of a seal in figure 3.[5] The Schlumberger Industries LVDTs provide a reliable method for precision measurement of linear displacement in the direction of the wall movement, perpendicular to the plane of the seal or overcast wall or deck. The LVDT is calibrated by varying the position of the core (the thin rod extending out from the cylindrical housing in figure 3) by known distances and then measuring the corresponding output voltages. Each LVDT was attached to an aluminum housing that was clamped to a steel post behind the seal, as shown in

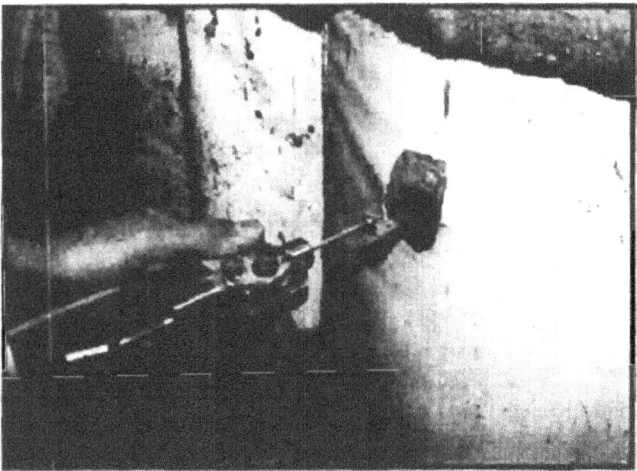

Figure 3.—LVDT attached to a seal.

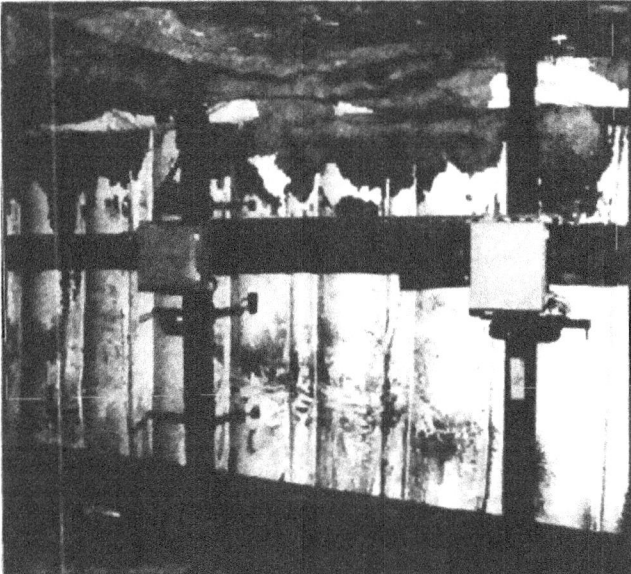

Figure 4.—Support posts and instrumentation on the back side of a seal.

figure 4. The square cross-section posts were bolted to the roof and floor. The main cylindrical body of each LVDT was held by the aluminum block, as shown held by the engineer in figure 3. The movable thin rod extending from the LVDT was attached to a small plate that was epoxied to the back face of the seal. These sensors were then interfaced to the nearest data-gathering station.

During the first test of the Barclay Mowlem program, the seal in crosscut 2 was instrumented with four LVDTs on the B-drift side (side of the seal opposite to the explosion). One LVDT was installed at the exact center (midheight and midwidth) of the seal (referred to as "middle" in the tables in appendix D). This is the sensor that is slightly below and to the right of the instrumentation box on the left post in figure 4. A second LVDT was installed at a three-quarter height and midwidth point (above and to the right of the left instrumentation box in

---
[5] All photographs in this report were taken by Eric S. Weiss, Kenneth L. Cashdollar, or William A. Slivensky of the NIOSH Pittsburgh Research Laboratory.

figure 4). (This is referred to as "upper" in the tables in appendix D.) A third LVDT was installed at a one-quarter height and midwidth point on the left post (referred to as "bottom" in the tables in appendix D). The fourth LVDT was installed at midheight and quarter-width (halfway between the seal center and the outby rib, just below the instrumentation box on the right post in figure 4). No sensors were installed on any of the stoppings since they were designed to release and vent the explosion overpressures. The overcast was instrumented with LVDTs (in a pattern similar to that used for the seals) on the outby B-drift side wall. Three LVDTs were also used on the overcast deck to measure the displacement of the deck. One LVDT was used for one of the overcast wing walls. Details of the instrumentation for the overcast are in the "Overcast" section for the Australian program found later in this report. For stronger explosion tests in which the seals and overcast had the potential to fail, some or all of the expensive LVDT sensors were removed so that they would not be destroyed.

The data gathered during the explosion tests were relayed from each of the data-gathering stations to an underground instrument room off of C-drift and then to an outside control building. A high-speed, 64-channel, PC-based computer data acquisition system was used to collect and analyze the data. This system collected the sensor data at a rate of 1,500 samples per channel over a 5-s period. The data were then processed using LabView, Excel, and PSI-Plot software and outputted in graphic and tabular form, as will be shown and discussed in the "Explosion and Air Leakage Test Results" sections for each of the three programs found later in this report. The reported data were averaged over 10 ms (15-point smoothing). This PC data analysis system allowed the data traces to be expanded in time and pressure (or other sensor value) so that the peak values could be read and recorded precisely.

AIR LEAKAGE DETERMINATIONS

An important factor to be considered for any seal design is its ability to minimize air leakage through the seal. Measurements of air leakages across the seals, stoppings, and overcast were taken before and after each of the explosion tests. For these air leakage tests, the D-drift bulkhead door (see figure 1) was closed. This directed all of the ventilation flow (from a vertical air-shaft in E-drift) toward C-drift. A wooden framework with brattice cloth or curtain was erected across C-drift outby the last seal or stopping position (figure 5). This curtain effectively blocked the ventilation flow, which resulted in a pressurized area on the C-drift side of the seals, stoppings, and overcast. By increasing the speed of the four-level LLEM main ventilation fan while in the blowing mode, the resultant pressure exerted on the structures increased from about 0.25 kPa (1-in H_2O) for the lowest fan speed setting to nearly 1.0 kPa (3.7-in H_2O) for the highest fan speed setting.

On the B-drift side of each seal and stopping design, a diaphragm of brattice cloth was installed across each crosscut (figure 5). A typical brattice with a 465-cm^2 opening near the

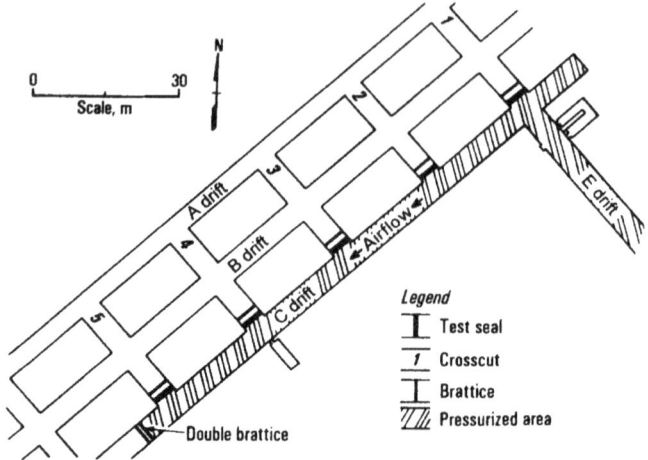

Figure 5.—Pressurized entry for leakage determination rates across the seals.

center is shown in figure 6. For the Barclay Mowlem overcast air leakage tests, three diaphragms were constructed: one inby the overcast in B-drift, one outby the overcast in B-drift, and one in the crosscut between A- and B-drifts. The 465-cm^2 center opening was only on one of these brattices. A vane anemometer was used to monitor the airflow through these openings on the brattices to determine the leakage rates through each design (figure 6). During construction of the seals, stoppings, and overcast, a copper tube was positioned through each of these structures with one end of the tube extending out on either side. This tube served as a way of measuring the air pressure exerted by the fan on each structure. During these air leakage tests, a pressure gauge was attached to the copper tube on the B-drift side to monitor the differential pressure across the structure.

As the ventilation fan speed was increased, the pressures on and the airflows through each structure were recorded. Based on data previously collected during the testing program with solid-concrete-block and cementitious foam seals [Stephan 1990a; Greninger et al. 1991], U.S. guidelines for acceptable air leakage rates through seals were developed for the LLEM seal evaluation programs. The air leakage rates through the seals during both pre- and postexplosion leakage tests were evaluated against these established guidelines. Table 2 lists these maximum acceptable air leakage rates as a function of pressure differential. For pressure differentials up to 0.25 kPa (1-in H_2O), air leakage through the seal must not exceed 2.8 m^3/min (100 cfm). For pressure differentials over 0.75 kPa (3-in H_2O), air leakage must not exceed 7.1 m^3/min (250 cfm). The flow rate was calculated from the linear air speed measured by the vane anemometer and the area of the opening through the brattice cloth behind each seal.

When a postexplosion visual inspection of any seal revealed substantial structural damage, that seal was considered not to meet the minimum standards as specified in the CFR for an underground coal mine seal and therefore "failed."

Figure 6.—Brattice in place for seal leakage test.

Postexplosion air leakage tests were not done on seals that showed significant damage, such as large, gaping cracks. The designs that withstood the pressure pulse with little or no outward signs of damage were tested for air leakage resistance. Postexplosion air leakage tests were not performed against the two Barclay Mowlem stopping designs since these stoppings were designed to partially vent the explosion pressure.

Table 2.—Guidelines for air leakage through a seal

Pressure differential		Air leakage rate	
kPa	in H$_2$O	m^3/min	cfm
<0.25	<1.0	<2.8	<100
0.25 < 0.50	1.0 < 2.0	<4.3	<150
0.50 < 0.75	2.0 < 3.0	<5.7	<200
>0.75	>3.0	<7.1	<250

CEMENTITIOUS PUMPABLE PLUG SEALS

Since 1990, several research programs have been conducted at the LLEM to evaluate the strength characteristics and air leakage resistance of pumpable cementitious plug seals [Stephan 1990a; Greninger et al. 1991; Weiss et al. 1993c; 1996; 1999]. These types of plug seal designs are not required to be hitched or keyed into the mine ribs and floor. Test results from those programs have shown that for a pumpable cementitious seal design to be deemed suitable by MSHA for use in an underground coal mine, that seal must be at least 1.2-m (4-ft) thick with a minimum grout compressive strength of 1.4 MPa (200 psi). Several seals using either cement foams [Stephan 1990a; Greninger et al. 1991] or cellular concretes [Weiss et al. 1996] are currently in use today in coal mines as a direct result of these LLEM seal evaluation programs.

In 1997, HeiTech Corp. entered into an agreement with PRL to evaluate four new cementitious pumpable seal designs. Under the agreement, HeiTech reimbursed NIOSH for all expenses incurred by NIOSH during this program. The seals were installed by HeiTech personnel. The following two sections discuss the construction process and the performance of these seals when subjected to a 138-kPa explosion pressure pulse.

CONSTRUCTION

As with the previously evaluated pumpable cementitious seal designs, these new HeiTech designs used a similar wooden framework and brattice liner to contain the cementitious slurry. Before installing these form walls, the concrete mine floor of the LLEM was roughened. (In a coal mine installation, the mine floor must be cleaned to the solid.) All loose material was removed from the LLEM roof, ribs, and floor. No hitching or keying of the seal is required with pumpable cementitious seal designs. The upright posts of the walls consisted of 15- by 15-cm (6- by 6-in) rough-cut posts wedged at the floor and roof (figure 7). The post pattern required a 76-cm spacing, with a maximum spacing not to exceed 91 cm. The end posts of each form wall are set as close to the rib as possible. The front form wall was not tied into the back form wall, except for the aggregate-grout seal in crosscut 2, which used two sections of 8-mm (5/16-in) diameter chain for each front/back post set, with a chain spacing of 60 cm and 140 cm from the mine floor. The spacings between the front and back form walls for each seal design were as follows: 865 mm (34 in) for the seal in the second crosscut, 915 mm (36 in) for the third crosscut, 760 mm (30 in) for the fourth crosscut, and 610 mm (24 in) for the fifth crosscut. A high-strength pumpable cementitious seal design that had been successfully evaluated during a previous program was still located in the first crosscut. Figure 7 shows the construction of the wood and brattice cloth form walls used for the seal in crosscut 4, with the brattice on the back wall but not yet on the front wall. Horizontal support boards consisted of 2.5-cm by 15-cm (1-in by 6-in) rough-cut lumber. To complete the form walls, these boards were attached across the front form wall and the back form wall upright posts using nails. The bottom horizontal board of each form wall rested on the mine floor and was cut to closely match the rib contours. The top horizontal board on the back form wall was anchored tight to the mine roof. The top horizontal board on the front form wall was anchored about 5 cm from the mine roof to allow for the installation of the bleeder ports. The remaining horizontal support boards on the front and back form walls were attached to the upright posts with a spacing of about 10 cm for the seal in crosscut 3 and about 76 cm for the seals in crosscuts 2, 4, and 5. Additionally, 5- by 5-cm square wire meshing with a 3-mm (1/8-in) diameter wire was attached to the inside of the front and back form walls for the seals in crosscuts 2, 4, and 5. No wire mesh was on the form walls for the seal in crosscut 3. The brattice liner covered the inside front and back form walls with a 15-cm overlap to the inside mine surfaces.

The cementitious slurry is mixed and pumped into each seal using a mixer-type placer pump. For each seal design, a three-port injection process was used during the pumping of the final slurry. These injection ports, with equal horizontal spacings, are installed into the top of the front form wall and angled to the mine roof to ensure uniform distribution of the cementitious slurry (figure 8). Two or three bleeder tubes installed near the mine roof were used during the final slurry injection for the seals in crosscuts 2, 3, and 5. No bleeder tubes were used on the seal in crosscut 4. Bleeder tubes provide a reliable method for determining when the cementitious slurry reaches the mine roof. These tubes are equally spaced along the seal and installed so as to ensure slurry contact within the highest roof cavity areas located between the form walls. Table 3 summarizes the construction schedule for the seals in crosscuts 2 through 5.

Duplicate samples of the cementitious grout were taken for each seal at various intervals during the slurry injection process. One-half of the samples were tested for compressive strength by MSHA; the rest were tested by an independent lab. During seal construction, the LLEM temperature varied from 9 °C to 14 °C (48 °F to 57 °F); the relative humidity varied from 57% to 98%.

The design in crosscut 2 consisted of a 865-mm (34-in) thick reinforced seal using aggregate (1- to 2-cm limestone) and HeiTech's hydrocrete cementitious material. Summary data on the size of this and the other seals are in table 4. Hydrocrete, made by Blue Circle Special Cements, Barnstone, U.K., is designed with a water-to-hydrocrete ratio of 1.5:1. The combined ideal density of the aggregate and grout for this seal design is 2,000 kg/m³ (125 lb/ft³). For additional anchoring of this seal, standard-grade roof bolts were installed along the centerline of the seal into the ribs, roof, and floor, with two along each rib and three each into the roof and floor. About one-half of each 1.8-m-long bolt was anchored within the roof hole using resin; the remaining 0.9 m extended into the seal. On average, the rib bolts were embedded about 0.7 m; the rest of the bolt extended into the seal. The floor bolts were embedded and grouted into the floor to a depth of about 0.8 m; the rest of the bolt extended into the seal. Next, the front and back form walls were installed. An aggregate was stowed between the two walls

Figure 7.—Construction of the wood and brattice cloth form walls used to contain the pumpable cementitious grout slurry.

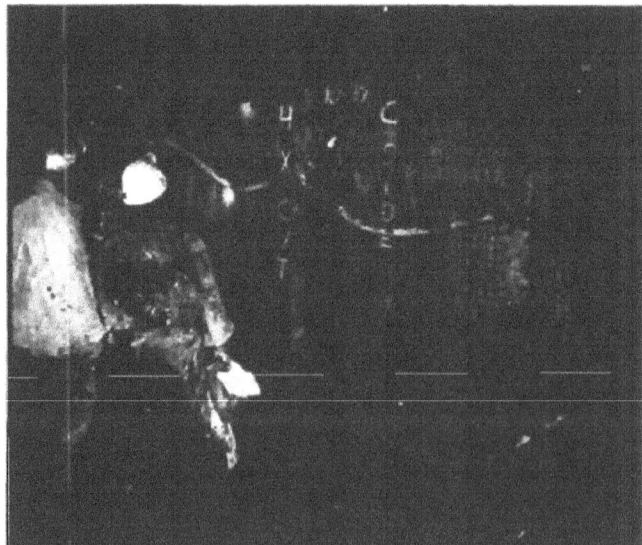

Figure 8.—Slurry injection using the three injection ports located near the mine roof.

Table 3.—Construction schedule at the Lake Lynn Experimental Mine

Design/type	Position	Grout injection	Shotcrete and final installation	Removal
HEITECH PROGRAM				
Seal, hydrocrete	Crosscut 2	Oct. 17, 1997	—	—
Seal, ribfill	Crosscut 3	Oct. 15, 1997	—	—
Seal, hydroseal	Crosscut 4	Oct. 16, 1997	—	—
Seal, hydroseal	Crosscut 5	Oct. 16, 1997	—	—
BARCLAY MOWLEM PROGRAM				
Seal	Crosscut 2	Feb. 2-5, 1998	Feb. 6-7, 1998	—
Stopping, water-filled	Crosscut 3	—	Feb. 7, 1998	After No. 358.
Stopping, air-filled	Crosscut 4	—	Feb. 11, 1998	After No. 358.
Seal	Crosscut 3, high roof.	Feb. 17-18, 1998	Feb. 20, 23, 1998	After No. 360.
Overcast	Intersection of crosscut 3 and B-drift.	Mar. 7, 9-11, 14, 16, 18-19, 1998.	Mar. 11, 13, 17-21, 23-24, 1998.	—
New seal	Crosscut 4	Mar. 24, 1998	Mar. 24-25, 1998	—
New seal	Crosscut 3, high roof.	Apr. 1, 1998	Apr. 2, 1998	—
PACKSETTER SEALS WITH SOLID-CONCRETE-BLOCK				
Seal, mortared	Crosscut 2	May 21, 1998[1]	—	—
Seal, mortared	Crosscut 3	May 21, 1998[1]	—	—
Seal, dry-stacked	Crosscut 4	May 19, 1998[1]	May 20, 1998[2]	—

[1]Grout filling of Packsetter bags.
[2]Sealant application to block wall.

Table 4.—Seals and stoppings size data

Design/type	Position	Thickness mm	Thickness in	Width, m	Height, m	Area, m^2
HEITECH PROGRAM						
Seal, hydrocrete	Crosscut 2	865	34	5.9	2.1	12.4
Seal, ribfill	Crosscut 3	915	36	5.9	2.1	12.4
Seal, hydroseal	Crosscut 4	760	30	5.8	2.3	13.3
Seal, hydroseal	Crosscut 5	610	24	6.0	2.2	13.2
BARCLAY MOWLEM PROGRAM						
Stopping, water-filled	Crosscut 3	170	7	5.9	2.1	12.4
Stopping, air-filled	Crosscut 4	300	12	5.8	2.3	13.3
Seal	Crosscut 2	[1]450	18	5.8	2.1	12.2
Seal	Crosscut 3, high roof	[1]450	18	5.9	2.8	16.5
New seal	Crosscut 3, high roof	[1]240	10	5.9	2.8	16.5
New seal	Crosscut 4	[1]165	7	5.8	2.3	13.3
PACKSETTER SEALS WITH SOLID-CONCRETE-BLOCK						
Seal, mortared	Crosscut 2	[2]405	16	5.8	2.1	12.2
Seal, mortared	Crosscut 3	[2]405	16	5.9	2.1	12.4
Seal, dry-stack	Crosscut 4	[2]405	16	5.8	2.2	12.8

[1]In addition, a 25-mm-thick coating of gunite was applied to C-drift side of each seal.
[2] ~80-cm by ~40-cm center pilaster.

to a depth of ~76 cm. The cementitious slurry was then pumped through a ~15-m-long, 30-mm-diam hose to the seal location until the slurry filled the void spaces between the aggregate. Once the slurry level reached the top of the first aggregate lift, a 36-cm-high second lift of aggregate was stowed, followed by subsequent slurry injection. The third lift was similar to the second, and the final lift of 71 cm completed the seal. About 12.5 t of aggregate and 157 bags, or 3,920 kg, of hydrocrete were used to build this seal. Compressive strength tests of the samples done by MSHA averaged ~4.1±2.1 MPa (~600 psi); those by an independent lab, ~5.8±3.4 MPa (~840 psi). The differences in the compressive strength test results are, in part, due to the continuous injection process used during the construction of these seals. This means that the dry powder and water are continuously mixed as the seal is poured. This leads to more variability than a batch-mixing process where the exact amounts of dry powder and water can be blended together thoroughly before injection. Although there is considerable variation in the compressive strength data, the data from MSHA and the independent lab agreed to within the standard deviations.

The design in crosscut 3 consisted of a 915-mm (36-in) thick seal using HeiTech's ribfill cementitious material. Ribfill, manufactured by Blue Circle Special Cements, uses a water-to-grout ratio of 2:1 with an ideal density of 1,200-1,280 kg/m^3 (75-80 lb/ft^3). As with the other seals, the ribfill was injected between the form walls and the final material was pumped using the three-port injection process (with bleeder tubes). Maximum recommended pumping distances should not exceed 365 m (1,200 ft). For this seal, ~90 m of 30-mm-diam hose was used during the LLEM installation. A total of 212 bags, or ~5,445 kg of material, was used for this seal. Compressive strength tests of the samples tested by MSHA averaged ~2.8±0.6 MPa (~400 psi); those by an independent lab, ~3.3±1.1 MPa (~480 psi). Figure 9 shows the completed seal in crosscut 3.

The design in crosscut 4 consisted of a 760-mm (30-in) thick seal, and the design in crosscut 5 consisted of a 610-mm (24-in) thick seal; both used HeiTech's hydroseal cementitious material. Hydroseal, also manufactured by Blue Circle Special Cements, uses a water-to-grout ratio of 1.2:1, with an ideal density of ~1,440 kg/m^3 (~90 lb/ft^3). The form walls and slurry injection process were similar to those used for the other seals, except that no bleeder ports were used during the final slurry injection period for the seal in crosscut 4. For this slurry, maximum recommended pumping distances should not exceed 300 m, and ~90 m of 30-mm-diam hose was used during the LLEM installation of seals 4 and 5. A total of 217 bags, or ~5,445 kg of material, was used for the 760-mm-thick seal in crosscut 4. A total of 149 bags, or ~3,810 kg of material, was used for the 610-mm-thick seal in crosscut 5. For the 760-mm-thick crosscut 4 seal, the compressive strength tests of the samples tested by MSHA averaged ~3.9±0.6 MPa (~565 psi); those by an independent lab, ~2.7±1.2 MPa (~390 psi). For the 610-mm-thick crosscut 5 seal, the compressive strength tests of the sample cylinders tested by MSHA averaged ~5.2±1.9 MPa (~750 psi); those by an independent lab, ~4.1±1.9 MPa (~600 psi).

Figure 9.—Completed ribfill seal in crosscut 3.

EXPLOSION AND AIR LEAKAGE TEST RESULTS

Air leakage tests were conducted against the four seal designs before conducting the first explosion test. As listed in table A-1 of appendix A, the preexplosion air leakage rates through each of the four seal designs were well within the established guidelines (see table 2).

The first explosion test (test 354 in table 1) generated a pressure pulse ranging from 190 kPa (28 psi) at the seal in crosscut 2 to 140 kPa (20 psi) at the seal in crosscut 5. The detailed listing of maximum static pressures at the data-gathering stations and at the seals for this explosion test is in table B-1 of appendix B. The maximum pressure at each of the seals and the summary result for each seal for explosion test 354 and the other tests are in table 5. For explosions with pressures >140 kPa (20 psi), the pressures in table 5 and appendix B are rounded to the nearest 5 kPa and to the nearest 1 psi. As discussed previously in the "Instrumentation" section of this report, these maximum pressure values in both table 5 and appendix B were smoothed over a 10-ms time period.

No damage was observed on the 865-mm-thick hydrocrete/ aggregate seal in crosscut 2 after being subjected to the 190-kPa static pressure pulse. Also, no damage was evident on the 760-mm-thick hydroseal in crosscut 4 (155-kPa (23-psi) pressure) or the 610-mm-thick hydroseal in crosscut 5 (135-kPa (20-psi) pressure). Postexplosion air leakage rates (table A-2 of appendix A) for the seals in crosscuts 2, 4, and 5 were well within the guidelines. For the 915-mm-thick ribfill seal in crosscut 3, minor damage occurred along the seal interface with the mine roof after being subjected to a static pressure pulse of 165 kPa (24 psi). Several horizontal and vertical cracks were also observed on the C-drift (explosion) side of this seal where the explosion forces had pulled the brattice away from the seal (figure 10). Postexplosion air leakage rates for seal 3 (table A-2) were well in excess of the established guidelines. After the postexplosion air leakage test, the front and back form walls and brattice were removed to allow a more comprehensive examination of this seal. It was determined that the excessive postexplosion air leakage rates were caused mainly by the large gaps between the top level of the ribfill and the mine roof. Improper placement of the three injection ports on the front form wall along the mine roof resulted in inadequate contact of the ribfill grout with the mine roof. The pressure pulse generated from the explosion damaged the brattice contact with the mine roof, exposing the gap between the ribfill grout and the mine roof; this resulted in the excess air leakage rates.

A second similar explosion test (LLEM test 355) was conducted against the four seal designs. All of the form walls and brattice were removed from both sides of each seal before this second test. The intent was to establish that the form walls were not significantly contributing to the overall strength of any of these seals, but instead were acting mainly as containers while the cementitious grout was poured and while it cured. Before conducting another preexplosion air leakage test (required since the form walls were removed from each seal), the gaps between

Table 5.—Evaluations of the seal, stopping, and overcast designs

Test No.	Date	Static pressure psi	Static pressure kPa	Evaluation	Result
				HEITECH PROGRAM	
354	Nov. 6, 1997	28	190	Crosscut 2 hydrocrete/aggregate seal	Survived.
		24	165	Crosscut 3 ribfill seal	Survived.[3]
		23	155	Crosscut 4 hydroseal seal	Survived.
		20	140	Crosscut 5 hydroseal seal	Survived.
355	Nov. 20, 1997	27	190	Crosscut 2 hydrocrete/aggregate seal	Survived.
		24	165	Crosscut 3 ribfill seal	Survived.[2]
		22	150	Crosscut 4 hydroseal seal	Survived.
		18	125	Crosscut 5 hydroseal seal	Survived.
				BARCLAY MOWLEM PROGRAM	
358	Feb. 11, 1998	4.0	27	Crosscut 2 seal	Survived.
		2.8	19	Crosscut 3 stopping, water-filled	NA.
		2.0	14	Crosscut 4 stopping, air-filled	NA.
359	Feb. 27, 1998	30	205	Crosscut 2 seal	Survived.[3]
		25	170	Crosscut 3 seal	Survived.[3]
360	Mar. 3, 1998	54	370	Crosscut 2 seal	Survived.[3]
		69	475	Crosscut 3 seal	Failed.
361	Mar. 26, 1998	[1]2.3	[1]16	Overcast	Survived.
362	Mar. 31, 1998	[1]4.3	[1]30	Overcast	Survived.
363	Apr. 1, 1998	[1]6.8	[1]47	Overcast	Survived.
364	Apr. 3, 1998	28	195	Crosscut 2 seal	Survived.
		23	160	New crosscut 3 seal, 24-hr	Failed.
		17	115	New crosscut 4 seal	Failed.
		[1]6.0	[1]41	Overcast	Survived.
				PACKSETTER SEALS WITH SOLID-CONCRETE-BLOCK	
365	June 22, 1998	22	150	Crosscut 2 seal, mortared	Survived.
		19	130	Crosscut 3 seal, mortared	Survived.
		18	120	Crosscut 4 seal, dry-stacked	Survived.[3]
366	June 25, 1998	27	185	Crosscut 2 seal, mortared	Survived.
		22	155	Crosscut 3 seal, mortared	Survived.
		18	125	Crosscut 4 seal, dry-stacked	Failed.

[1]Static pressure in crosscut 3, leading to overcast.
[2]Sealant reapplied prior to explosion test.
[3]Physically survived explosion, but failed U.S. air leakage test.

Figure 10.—Horizontal cracks evident near the mine roof on the ribfill seal in crosscut 3 after test 354. Large gaps between the mine roof and the top level of the cured ribfill are evident where incomplete closure was obtained during the initial grout injection process.

the ribfill grout and the mine roof were filled with ribfill grout for the seal in crosscut 3. The results of this second pre-explosion air leakage test (table A-3 in appendix A) showed that all four seal designs were well within the established guidelines. The second explosion test subjected the seals to the following peak static pressures (details are in table B-2 and summarized in table 5): 190 kPa (27 psi) for the 860-mm-thick aggregate/hydrocrete design in crosscut 2, 165 kPa (24 psi) for the 915-mm-thick ribfill design in crosscut 3, 150 kPa (22 psi) for the 760-mm-thick hydroseal design in crosscut 4, and 125 kPa (18 psi) for the 610-mm-thick hydroseal design in crosscut 5. The unsmoothed pressure data interpolated for seal 5 showed peaks >138 kPa. Observations of the seals after the second explosion test (LLEM test 355) revealed no apparent damage to any of the four seals. Postexplosion air leakage measurements (table A-4) were within the established guidelines for all four seals.

Based on these results, MSHA has determined that these four HeiTech seal designs are suitable for use in underground U.S. coal mines with the following restrictions. These seal designs as constructed and tested in the LLEM cannot be used in entries with roof heights >2.4 m or with entry widths >6.1 m without design modifications and prior review of the written ventilation plan by the MSHA District Manager. This particular height and width requirement applies to all other seal types as well, e.g., concrete block, wood block, polymer, etc., that are proposed for use in a mine's written ventilation plan. The wood and brattice form work for these seal designs is not considered part of the seal. However, if the form work is removed or any part of the seal grout material becomes exposed, it is necessary to coat the material with an MSHA-approved sealant [Sawyer 1992]. MSHA has additional detailed specifications for the average and minimum compressive strengths of the samples collected during seal installation for these four seal designs. The results of this research program showed that thinner pumpable cementitious plug seals could withstand a 138-kPa (20-psi) explosion if they had higher compressive strengths. Previously tested pumpable cementitious plug seals with a compressive strength of 1.4 MPa (200 psi) had to be at least 1.2 m (4 ft) thick [Stephan 1990a; Greninger et al. 1991; Weiss et al. 1993c, 1996]. The results of LLEM tests 354 and 355 showed that plug seals that are 0.6 to 0.9 m thick could withstand a 138-kPa explosion if their average compressive strengths were at least 4.7 and 3.0 MPa (680 and 435 psi), respectively. Even thinner (0.3-m) pumpable cementitious seals with much higher compressive strengths of ~40 MPa and additional anchoring had also withstood a 138-kPa explosion in a previous program [Weiss et al. 1999].

AUSTRALIAN DESIGN SEALS, STOPPINGS, AND OVERCAST

A particular hazard in gassy underground coal mines occurs when a section of the workings is sealed because of the effect of spontaneous combustion. If methane is being continually generated, the atmosphere behind the seals could enter the explosive range for methane-air mixtures in a fairly short period of time, and spontaneous combustion could provide an ignition source. Under these circumstances, an explosion could occur 1-2 days after the seal is completed. On August 7, 1994, 11 miners and 1 contractor were killed when a methane-air mixture ignited within a recently sealed room-and-pillar panel at the BHP Australia Coal Moura No. 2 coal mine in Queensland, Australia [Roxborough 1997]. The most likely ignition source was determined to be the heating caused by spontaneous combustion within the sealed area. The overpressures generated from the methane ignition resulted in the failure of several seals that were newly installed about 22 hr before the ignition. As a result of this disaster, a considerable public outcry demanded that an in-depth inquiry be conducted to determine the cause of the explosion and to recommend ways to prevent future occurrences in the Queensland coal mines.

In late 1997 and early 1998, PRL collaborated on a joint research project with Barclay Mowlem Construction Ltd. of Queensland, Australia, to investigate the capability of various seal and stopping designs and an overcast design to meet or exceed the requirements of the Queensland Department of Mines and Energy's [1996] Approved Standard for Ventilation Control Devices. This standard was the result of deliberations and investigations by Task Group 5, which was formed by the recommendation of the Warden's Inquiry concerning the Moura No. 2 mine explosion [Roxborough 1997]. Task Group 5 was charged with the reassessment of the regulatory provisions for explosion-resistant seals and the investigation of mine inerting techniques. Similar to an evaluation program done in 1997 with Tecrete Industries Pty. Ltd. of New South Wales, Australia [Weiss et al. 1999], this evaluation program with Barclay Mowlem Construction Ltd. at the LLEM tested designs within a range of overpressures to match the recommendations of Task Group 5. The overpressure ratings for underground ventilation control devices in Australia are as follows: 14, 35, 140, and 345 kPa (2, 5, 20, and 50 psi). The expected outcome of the new standard for seals and airlocks in Queensland is that all ventilation control structures will have an overpressure rating based on an assessment of the risk and purpose of the particular control structure. These standards do not address the structural design or the material to be used in seal construction. The Barclay Mowlem seal designs would also be evaluated relative to the U.S. static pressure and air leakage requirements.

During the Barclay Mowlem research program, several seal and stopping designs and an overcast design were subjected to various explosion overpressures in the LLEM. As part of the evaluation, seals were built in crosscuts (cut-throughs) ranging from 2.1 to 2.8 m high. The higher roof had been enlarged previously, forming a roadway with dimensions that are representative of those found in Australian underground coal mines and some U.S. coal mines. As with the previous Australian program, one particular requirement of the Barclay Mowlem program was to test an isolating seal design that could withstand an explosion producing a static horizontal overpressure of ~140 kPa within 24 hr of its completion. Additionally, this Barclay Mowlem program was the first time that an overcast design had ever been evaluated for strength characteristics and air leakage resistance when subjected to full-scale explosion overpressures.

CONSTRUCTION OF SEALS, STOPPINGS, AND OVERCAST

Several Barclay Mowlem seal and stopping designs and an overcast design, using specially designed reinforced vinyl bladders in-filled with cementitious grout, were tested in the LLEM. There were two parts to the testing program. The first part involved building and testing two seal designs at explosion overpressures of >140 kPa (20 psi). Two stopping designs were also evaluated at explosion overpressures of 14-19 kPa (2-3 psi). The second part of the program was the testing of an overcast design and two additional seal designs.

The Barclay Mowlem seal and overcast designs all used a vinyl bladder system suspended by a metal framework anchored to the mine roof. The bladders were made with a double-sided white polymeric coated polyester fabric (VYNA 516, manufactured by Southcorp Industrial Textiles, Clayton, Victoria, Australia) with antistatic and fire-retardant characteristics. The various vinyl bladder configurations are discussed in the next section on "Seals." These vinyl bladders were preinflated with air before the grout-filling process. The portland cement-based grout, B M Mine Grout, was made in Australia by Blue Circle Southern Cement Ltd., Greystances, New South Wales, Australia. Compressive strength tests on the grout as conducted by Blue Circle Packaging in Australia provided the following results: 10 MPa (1,450 psi) at 1 day, 18 MPa (2,610 psi) at 2 days, 48 MPa (6,960 psi) at 7 days, and 68 MPa (9,860 psi) at 28 days. The grout was pumped into the suspended, preinflated vinyl bladders using a GP 2000 pump. This is a compressed air-driven modular unit that encompasses a mixing bowl mounted over a receiver hopper that has a screw feed to a small positive displacement pump. The prebagged dry grout mix was combined with a measured amount of water to achieve the correct material strength characteristics and then pneumatically pumped through a 50-mm-diam hose into the vinyl bladder. Shotcrete operations used the same prebagged shotcrete grout mix that was used inside the vinyl bladders of the seals and overcast. The shotcrete was spray applied using a Meyro Piccola machine. This shotcrete machine pneumatically delivers the dry shotcrete to a mixing nozzle where an operator can adjust the volume of water to achieve the desired consistency of the material applied. When handling these cementitious products, all safety data sheet instructions should be adhered to by the operators.

A summary of the construction data for the four seals, two stoppings, and overcast is found in tables 3 and 4. The grout injection dates and final shotcreting dates are listed in table 3. The last column of table 3 shows the LLEM explosion test after which the stopping or seal was removed. The dimensions of the stoppings and seals are listed in table 4. Additional construction details for the seals, stoppings, and overcast are found in the next three sections of this report. The designs were built in the LLEM under conditions similar to those that may be encountered during seal construction in an underground coal mine. As in the installation of any seal, all loose material had to be removed from the seal construction site, leaving competent strata. It must be noted that the test environment at the LLEM is one of solid, nonyielding strata. Previous practice during some seal evaluations in the LLEM was to provide edge restraint by bolting 150-mm by 150-mm steel angle (12 mm thick) to the floor and ribs. These steel angles were attached using 600-mm-long, 25-mm-diam grade 8 steel all-thread rod (embedded 450 mm) or 230-mm-long, 25-mm-diam Hilti Kwik bolt fasteners as manufactured by Hilti, Inc., of Tulsa, OK. Both rods and bolts used 450-mm spacings on the floor and rib. Several U.S. operating coal mines have been permitted to use a similar type of edge restraint in areas with hard sandstone floors in which standard keying would be very difficult. To achieve the desired hitching (recessing) for this program, the concrete floor was trenched to a depth of either 150 or 300 mm (6 or 12 in) and/or the 150-mm (6-in) steel angle was used. The hitching on the ribs was simulated by bolting the 150-mm by 150-mm by 12-mm-thick steel angle to each of the ribs on either side of the structure. The 25-mm-diam steel bolts were full-resin anchored and embedded 900 mm into the ribs on approximately 450-mm centers.

The mine air temperature during the 3-month construction and testing period ranged from 8 °C to 19 °C (47 °F to 66 °F) and averaged 11.5 °C (52.5 °F). The relative humidity ranged from 52% to 76.5% and averaged 66%. Heaters were used in the immediate vicinity of the seal and overcast locations to raise the mine temperature during the grouting operations. The grout used in this program was formulated for use in Australian coal mines, which are typically 5 °C to 10 °C (10 °F to 20 °F) warmer than U.S. mines. Initial compressive strength test results from the grout samples taken during the construction of the first seal showed that the cooler temperature of the LLEM slowed the cure time of this grout, thereby affecting the short-term strength characteristics of the grout, i.e., the compressive strengths of the grout samples as cured in the LLEM were typically lower for a given cure period compared with those obtained from samples taken in the warmer Australian mines. Given the time constraints of the program, the heaters were used to compensate for the cooler mine temperatures and allowed the grout to cure in temperatures typical of Australian coal mines.

Seals

During this Barclay Mowlem program, the seals were built with a cementitious-based grout with polyfibers. The grout was pumped into a vinyl bladder assembly. The vinyl bladder used for the first two seal designs and the overcast wall designs consisted of a series of interlocking tubes joined together by baffles (figure 11). The baffles provide a means to allow the grout to flow through the interlocked vinyl tubes. This results in a uniform distribution of the grout throughout the entire width of the seal. Two 25-mm-diam valves (shown as "bleeder ports" in figure 11*A*) are attached to the bladder as a means to inflate air before the grout injection process and for subsequent air release during the grout injection process. Two 50-mm-diam valves are attached to the bladder for the grout injection.

At the start of the seal installation, a spreader bar, which is built from 50-mm by 50-mm medium gauge tubing spaced 300 mm apart, is attached to the mine roof. During installation, the spreader bar assembly is held in position by roof jacks (figure 12). A flat metal strap is placed over the spreader bars and then bolted to the mine roof using 900-mm-long by 25-mm-diam fully encapsulated resin roof bolts on 2-m spacings. After the spreader bar assembly is installed, the roof jacks are removed and shotcrete is applied to the spreader bar assembly (figure 13). The vinyl bladder is pulled into place by ropes attached to a series of hooks on the spreader bar (figures 13-14) and then inflated. Between each of the inflated large vinyl bladder tubes are smaller 100-mm-diam piers (see figure 11*A*)

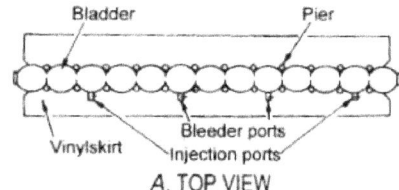
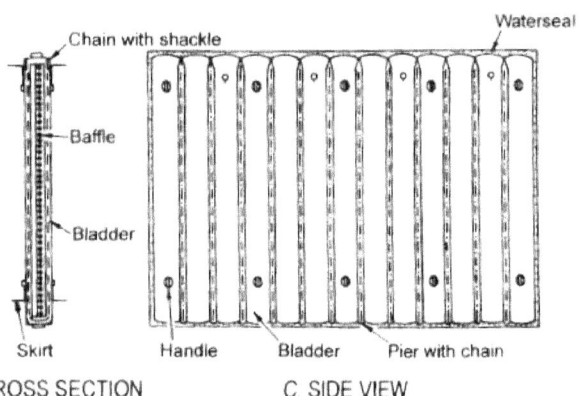

Figure 11.—Schematic of vinyl bladder with internal baffles used for construction of the seal and overcast designs. A, top view; B, cross section, C, side view.

Figure 12.—Spreader bar anchored to the mine roof used to support the seal bladder system.

Figure 13.—Shotcreting of the spreader bar and hook assembly.

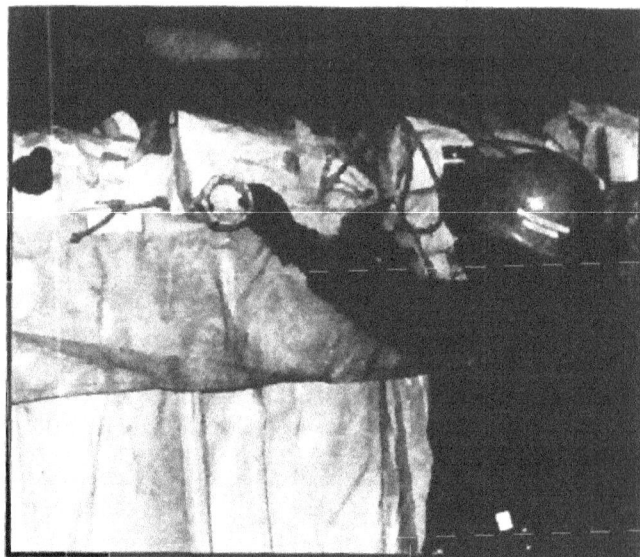

Figure 14.—Inflated vinyl bladder assembly showing the injection port for the piers.

that are interlocked to the outside and overlapped under the main bladder assembly. The piers are made from the same vinyl fabric as the bladder. For additional reinforcement, a light chain is attached to the spreader bar hooks and then dropped into each pier. At the bottom, the chains from the piers on either side of the seal are tucked under the main vinyl bladder. Figure 14 shows the vinyl bag attached to the spreader bars, with an additional person holding one of the injection ports.

Variations occurred during the installation of the seals in the LLEM compared to the standard installation in an Australian coal mine. This was due, in part, to the solid limestone strata, the concrete floor, and the entry geometry in the LLEM. Recessing of the seal into the roof and ribs was simulated using the 150-mm by 150-mm by 12-mm-thick black steel angle (standard coal mine installation would require recessing to 300 mm). Recessing of the seals into the floor was by a combination of trenching and/or steel angle. The concrete floor of the LLEM was trenched to a depth of 300 mm for the first two seal designs (installed in crosscuts 2 and 3); this was similar to the requirement for a standard seal installation in a coal mine. Steel angle was also used on the floor on the B-drift side of the crosscut 2 seal, providing an equivalent hitching depth of

450 mm. The steel angle was attached using 900-mm-long by 25-mm-diam full-resin bolts. Shotcrete was then sprayed along each edge between the steel angle and strata using the Meyco pneumatic shotcrete machine. The shotcrete sealing mix is the same product as the pumpable seal grout (Blue Circle's B M Mine Grout). Figure 15 shows the spreader bar assembly attached to the roof and the steel angles attached to the roof, ribs, and floor. There is also a metal strap across the back side of the seal at midheight.

After installation of the steel framework, the vinyl bladder was placed in position and inflated with air (figure 16). The bladder was then attached to the spreader bar using connecting chains that were looped over the spreader bar hooks. The piers on each side of the bladder were then shackled to these chains. After the support chain was dropped into the piers, it was also shackled to the spreader bar. The bottom of the vinyl pier and chains was then tucked under the main body of the vinyl bladder on each side. The grout was then pumped into each of the vinyl piers, filling each pier completely to the top. The grout-filled piers were then allowed to cure overnight. Next, the grout was pumped into the vinyl bladder in lifts of about 500 mm each. The baffles between the interlocked vinyl tubes allow the grout to flow and fill the entire bladder evenly. Before pumping the second lift of grout, the grout in the first lift was allowed to firm up, but not too firm as to cause a cold joint. The air within the bladder was released through the 25-mm-diam bleeder valves. Toward the end of the pumping operation, the pumping rate was varied to force all of the air and excess water from the bag. A series of small holes was inserted along the top of the bladder to relieve the remainder of the excess water and air. This process ensured that the grout-filled bladder was locked securely to the mine strata. The grout also has some expansion properties (1.5%), which serve to lock the seal into place.

Shotcrete was then sprayed around the edges of the seal to bond the seal to the mine strata. Depending on the seal height, two or three horizontal steel straps (see example of one such strap in figure 15) were bolted to each face (both front and back) of the seal using 150-mm-long by 25-mm-diam Hilti Dyna bolts. The 2.1-m-high seal in crosscut 2 had two straps on the C-drift face (figure 17) and one on the B-drift face. The 2.8-m-high seal in crosscut 3 had three straps on the C-drift side and two on the B-drift face. The straps were bolted to the mine ribs using 900-mm-long by 25-mm-diam full-resin bolts. A 25-mm-thick coating of shotcrete was then sprayed on the C-drift side of the seal to cover all exposed surfaces and on the perimeter of the seal on the B-drift side.

The 450-mm-thick seal in crosscut 2 was built in the manner described in the above paragraphs (see figures 15-17). The 450-mm-thick seal in the 2.8-m-high roof section of crosscut 3 was also built (figures 18-19) in a similar manner, except that the spreader bar was not recessed into the mine roof. Samples were taken during the grout injection process for compressive strength analyses at an independent lab. The compressive strength of the grout samples averaged 2.4 MPa (350 psi) after

Figure 15.— Framework for construction of seal in crosscut 2.

Figure 16.—Construction of seal in crosscut 2 showing vinyl bladder in place, but not yet filled.

Figure 17.—Completed seal in crosscut 2.

1 day, 24.6 MPa (3,570 psi) after 7 days, 33.4 MPa (4,840 psi) after 14 days, and 49.6 MPa (7,190 psi) after 28 days. These values were significantly lower than the original data measured by Blue Circle in Australia, perhaps because of the lower temperatures and the resulting longer cure times in the LLEM.

Two additional seals (the last two seals listed under the "Barclay Mowlem" section of tables 3 and 4) were built in crosscuts 3 and 4 toward the end of the evaluation program after the first seal was removed from crosscut 3. The seal in crosscut 4 (figure 20) was built just before the overcast evaluations and therefore had a 9-day cure period before being subjected to the explosion pressure pulse. The second seal in the high roof section of crosscut 3 (figure 21) was built after the overcast evaluations and was tested within ~24 hr of its completion. The seal in crosscut 3 consisted of a series of individual 240-mm-diam vinyl tubes connected to one another in modules of four (figure 21). The tubes used for the seal in crosscut 4 were 165 mm in diameter and likewise sewn together in modules of four (figure 20).

For both seals, the tube assemblies were suspended (figure 22) from chains from a spreader bar, which consisted of a 50-mm by 50-mm steel tubing assembly similar to that used with the other seals described earlier in this report. The tube assemblies were suspended from the spreader bar with 6-mm chains on 170-mm spacings for the seal in crosscut 4 and 240-mm spacings for the seal in crosscut 3. The spreader bar was not recessed into the mine roof and was attached using 1,800-mm-long by 25-mm-diam full-resin bolts on 2-m spacings. A heavy coating of shotcrete was sprayed onto the spreader bar and edges. A light chain was placed in each tube and connected to the spreader bar by shackles. The tube modules were open at the bottom. The vinyl material on the

Figure 18.—Construction of seal in high roof section of crosscut 3, showing the vinyl bladder being installed.

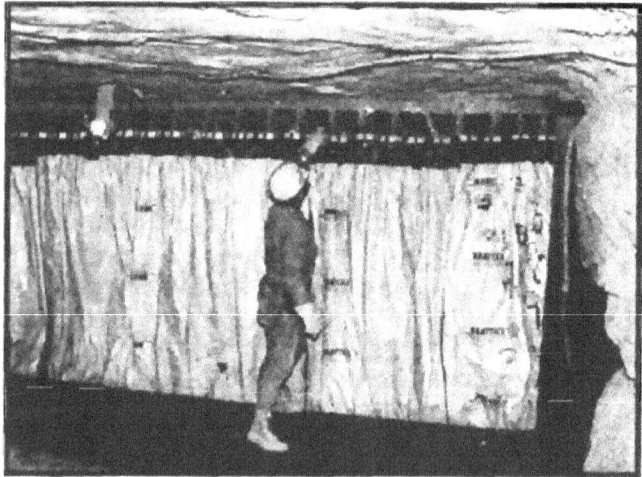

Figure 20.—Construction of seal in crosscut 4 showing the vinyl tubes before filling.

Figure 19.—Construction of seal in high roof section of crosscut 3, showing the grout injection hose attached to the bladder.

Figure 21.—Construction of second seal in high roof section of crosscut 3.

bottom of each module was simply folded over and fixed to the mine floor by placing steel strapping along the full length of the curtain. This strapping was bolted to the floor using 900-mm-long by 25-mm-diam full-resin bolts. It is important to keep the curtain gathered and baggy at the base to allow the tubes to form shape during filling. The attached vinyl side wings on the tube modules were folded back under the steel strapping and attached to the mine ribs by 1,800-mm-long by 25-mm-diam full-resin bolts. For each seal, the 150-mm by 150-mm by 12-mm-thick steel angle was used to simulate the rib recessing. A combination of trenching of the concrete floor and/or bolting of steel angle to the floor was used to simulate the floor hitching (recessing) for these two seals. The floor was trenched to a depth of 150 mm to provide hitching for the crosscut 3 seal, and the 150-mm steel angle was used to simulate hitching for the crosscut 4 seal.

The grout was pumped fully into each tube for each seal. The wire cables across the tops of the tubes in figure 22 provided additional reinforcement and helped to hold the grout in place. When the grout had set, three steel straps were attached to the C-drift seal face at the top, middle, and bottom of the seal in crosscut 3, and two steel straps were used on the C-drift face of the seal in crosscut 4. No straps were used on the B-drift face of either of these two seals. The straps were attached using 150-mm-long by 25-mm-diam Hilti Dyna bolts. These steel straps were then bolted to the mine ribs using 1,800-mm-long by 25-mm-diam full-resin bolts. A light wire mesh was attached across the top portion of each seal on the B-drift face and formed around each rib to provide some reinforcement and to serve as a backing along the top open section of the seals when applying shotcrete from the C-drift side. A 25-mm-thick coating of shotcrete was sprayed on the entire face and perimeter of the C-drift side (explosion side) of each seal. Shotcrete was then sprayed on the back (B-drift side) perimeter of each seal.

Stoppings

The construction dates, locations, and dimensions for the two stoppings are listed in table 3 and 4. The stopping in crosscut 3 is a unique design; it is composed of a number of 170-mm-diam heavy vinyl tubes that are joined together and filled with either stone dust or water (figure 23). Water was used during the LLEM program. The 170-mm-diam steel ring is welded or sewn onto the top of the vinyl tubes. Two holes are drilled through this ring to accommodate a shackle for attaching the tube to the spreader bar chains. The tubes are secured at the bottom and attached at the top by chains to the spreader bar. The spreader bar used for the stopping installation has a nominal size of 100 mm by 50 mm. Chains (6-mm) are welded to the spreader bar on 170-mm spacings and are used to suspend the individual tubes (figure 23). The spreader bar is attached to the mine roof using 900-mm-long by 25-mm-diam full-resin bolts on 2-m spacings. In a standard coal mine installation, the spreader bar would be recessed 150 mm into the roof, and the stopping would also be recessed 150 mm into the floor and 300 mm into the ribs. For the LLEM installation, the stopping was not recessed into the roof or floor. The 150-mm by 150-mm by 12-mm-thick steel angle was used to simulate the recessing into the ribs.

A Velcro strip (hook-and-latch fabric fastener) is sewn down each side of each of these tubes to allow each tube to be joined together to form a full-width curtain (figure 24). Plastic clips are also used to secure the individual tubes to one another. The floor seal is fabricated from two vinyl tubes sewn together on a vinyl pad. Between the two tubes along the entire length a Velcro strip was attached. These vinyl tubes are filled with stone dust in order to conform to the irregularities of the mine

Figure 22.—Construction of seal in crosscut 4 showing details of the tops of the vinyl tubes and light meshing overlay.

Figure 23.—Construction of water stopping in crosscut 3, with the individual tubes suspended from the roof-mounted spreader bar.

floor. This pad is anchored to the mine floor using 600-mm-long by 25-mm-diam full-resin bolts spaced 1 m apart. The Velcro strip attached to the bottom of each of the 170-mm-diam vinyl tubes attaches to its Velcro strip counterpart located on the floor pad to provide for a tight seal. A vinyl skirt is fixed at the top to the spreader bar and then fastened to the vinyl tubes using the Velcro fastening system. This provides for a tight seal for the top of the stopping. Similar vinyl sheets with Velcro fasteners are attached to the outermost tube on each side of the stopping and then attached to the mine rib. Shotcrete is sprayed along the top of the stopping covering the spreader bar and top vinyl curtain to completely seal this area. Shotcrete is then applied to the edges of the stopping, thus sealing the curtains to the mine ribs. Shotcrete is only applied to the bottom of the stopping on the positive-pressure side (C-drift side). This design is intended to allow the stopping to release, shed water (or stone dust) in the event of an explosion or overpressure, and then return to its original position, thereby restoring partial ventilation that would otherwise be lost.

The stopping design in crosscut 4 was developed as a temporary seal system that can be activated either from the mine outby side or remotely from the surface in the event of an emergency. It was designed to be converted into a permanent seal if necessary by replacing the air within the inflated bladder with grout. For the LLEM evaluations, this stopping design, referred to as a "quickseal," was only evaluated for its capability to withstand a low-level explosion without rupture of the inflated vinyl bladder. The spreader bar was not used because of unexpected complications during installation. It was decided to only evaluate the inflated bladder resistance to the low-level explosion pressure pulse. The steel rib angle was used to simulate a 150-mm recess on the C-drift ribs and floor. The contour of the mine roof and ribs on the B-drift side of the quickseal location provided recessing >150 mm. The bladder of the quickseal was not attached to the mine roof and was only held into position by the frictional forces exerted on the mine strata due to the internal pressure of the inflated bladder. No shotcrete was used.

A standard installation for the quickseal would require the spreader bar to be recessed high enough into the mine roof to accommodate the bladder and allow for a flush finish to the roof line. The spreader bar would be identical to that used for the seal designs and would be anchored to the mine roof in the same manner. To accommodate the deployed quickseal bladder, the mine floor would be recessed 300 mm and the mine ribs a minimum of 500 mm.

Overcast

The overcast design by Barclay Mowlem is a unique permanent ventilation system that is designed to provide a high degree of flexibility against ground movement, in addition to providing protection against low-level explosion pressures. This particular overcast was designed to provide a high load-bearing capacity required for the transport of mining machinery across the overcast deck.

The limestone mine roof at the intersection of B-drift and crosscut 3 was enlarged through explosive blasting to a maximum height of 6 m at the center and an enlarged area of ~35 m^2 in the intersection. The contour of the mine roof was then tapered inby and outby in B-drift to realign with the original height of the roof in B-drift; this occurred at ~3 m on either side of crosscut 3.

The overcast side walls form solid concrete walls across the B-drift entry and support the top overcast deck (figure 25). These side walls were built in a manner similar to the vinyl bag seal designs described in the earlier "Seals" section of this report. The 150-mm by 150-mm by 12-mm-thick steel angle

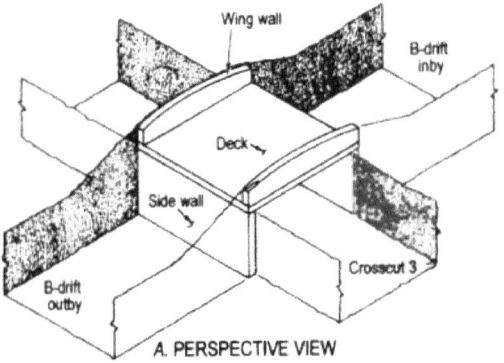

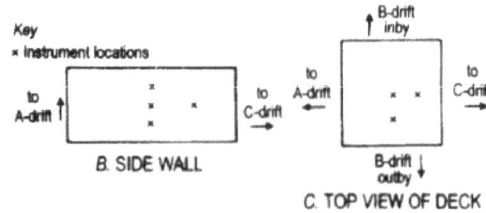

Figure 25.—Schematic drawing of overcast at the intersection of B-drift and crosscut 3. *A*, perspective view; *B*, side wall showing instrumentation positions, as viewed from B-drift outby; *C*, top view of deck showing instrumentation positions.

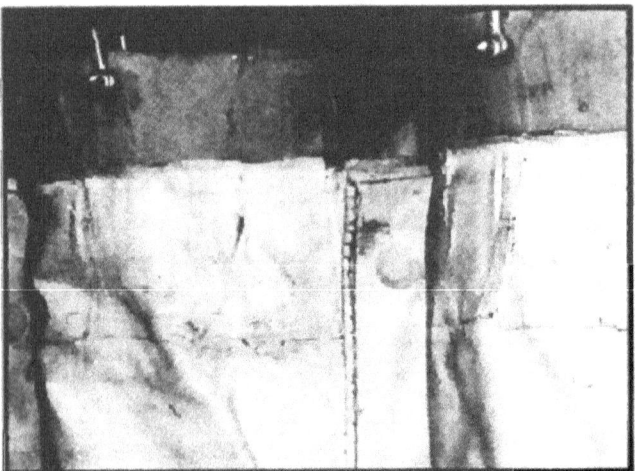

Figure 24.—Details of the Velcro and plastic clip fastening system for the water stopping in crosscut 3.

was used on the concrete floor to simulate a recess. This angle was bolted to the mine floor using 150-mm-long by 25-mm-diam Hilti Dyna bolts. The steel spreader bar used to suspend the vinyl bladder was attached to the top of a vertical steel angle framework that was prefabricated to achieve the desired height off of the floor: ~2 m for the LLEM installation (figure 26).

The 300-mm-thick vinyl bladders were attached to the spreader bar assembly and inflated with air. The connecting chains were looped over the spreader bar, and the 100-mm-diam vinyl piers were then shackled to the chain. One support chain was placed into each of the vinyl piers and shackled to the spreader bar. The bottom of each pier was tucked under the bladder on each side of the side wall. As with the seal construction, the vinyl piers were first filled with grout, and they were allowed to set overnight. Following this cure period for the grout in the piers, grout was then injected into the vinyl bladder in lifts of ~500 mm each. Each lift was allowed to firm up before the next lift, but not too firm as to cause a cold joint. The grout can be injected in either one of the two 50-mm-diam valves depending on the levels of the mine. This allows the grout to flow and fill the bladder evenly. Once the bag is filled, the pumping rate is decreased and then increased to force all excess air and water from the bag. This is done by inserting a series of holes along the top of the bladder. As all of the air and water are relieved, these small holes close off with grout, allowing the bladder to be locked securely into position.

After the grout injection process and overnight cure period, any grout and bag protruding above the spreader bar was removed to form a level surface for placement of the top deck. The main difference in the construction of this overcast design in the LLEM compared to a standard coal mine installation was that a short extension wall was required to be built at each end of both side walls in order to extend the side walls to and beyond the intersection corners (and away from the electrical and water connections embedded in the LLEM ribs). These extension walls are not shown in the figure 25 schematic, but one extension wall is shown on the far left side of figure 26. In a standard installation in an Australian coal mine, the side walls would be recessed into the ribs and would not require the extension walls. The extension walls were used because the intersection in the LLEM is larger than a typical Australian intersection. Steel angle was used to simulate the recess of these extension walls to the ribs. The 150-mm by 150-mm by 12-mm-thick steel angle was bolted to the ribs using three 900-mm-long by 25-mm-diam full-resin bolts. Shotcrete was sprayed around the rib and floor edges of the side and extension walls to completely bond the structure to the mine strata. A 25-mm-thick coating of shotcrete was also applied to both faces of these walls.

The top deck was composed of a steel Bondeck decking complete with side skirts. The thickness for the top deck (the height of the side skirts) depends on the particular use for that overcast, i.e., an overcast deck designed to allow machinery transport would be designed to contain more grout and thus would result in a thicker concrete slab compared to an overcast deck used only for ventilation control. The top deck panels were placed into position on top of the side walls and clipped together (figure 27). The deck assembly was then bolted down to the top of the side walls using 150-mm-long by 25-mm-diam Hilti Dyna bolts. A metal skirt was riveted around the perimeter of the Bondeck base to form an open top deck with a depth of 200 mm (figure 28). This 200-mm-thick top deck was designed to handle the transport of heavy mining machinery. Figure 29 shows the top of the deck, with steel rebar and steel mesh installed for reinforcement. Before pumping the grout, jacks were placed under the decking panels on 1-m spacings to support the weight of the slurry grout until the grout set (figure 30). Grout was pumped into the top deck form, filling it to the top of the metal skirt. After the grout had set, 900-mm-long by 25-mm-diam full-resin bolts were installed on 1.5-m spacings down through the 200-mm-thick top deck slab and into the top edge of the side wall.

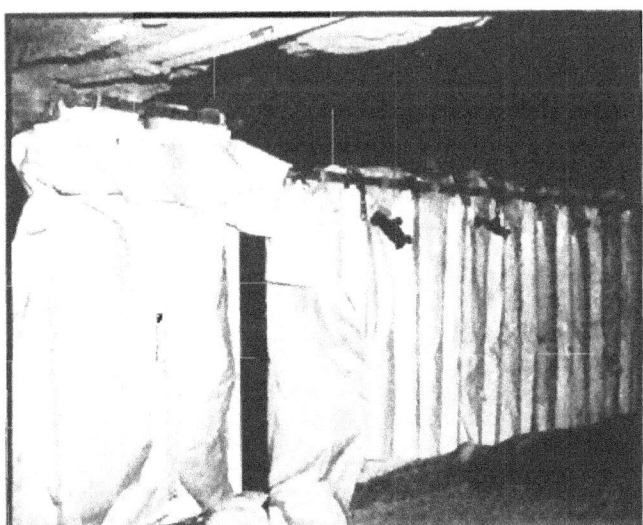

Figure 26.—Construction of side wall of overcast at the intersection of B-drift and crosscut 3.

Figure 27.—Construction of overcast at the intersection of B-drift and crosscut 3: installation of deck on top of side wall.

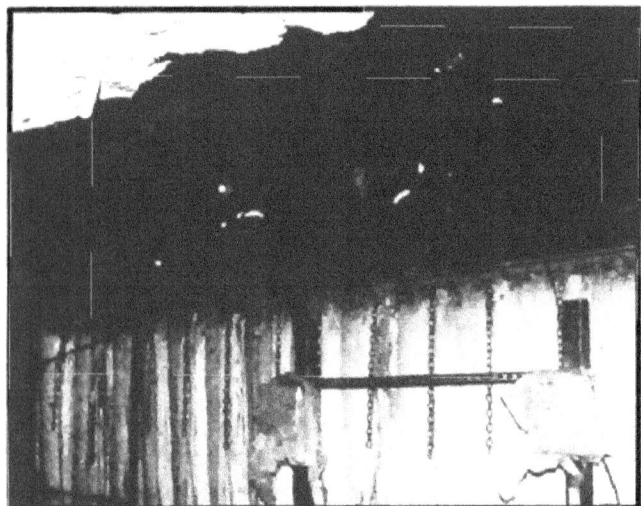

Figure 28.—Construction of overcast at the intersection of B-drift and crosscut 3: installation of skirt around edge of deck.

Figure 30.—View underneath the overcast deck showing temporary supports while the deck cement cured.

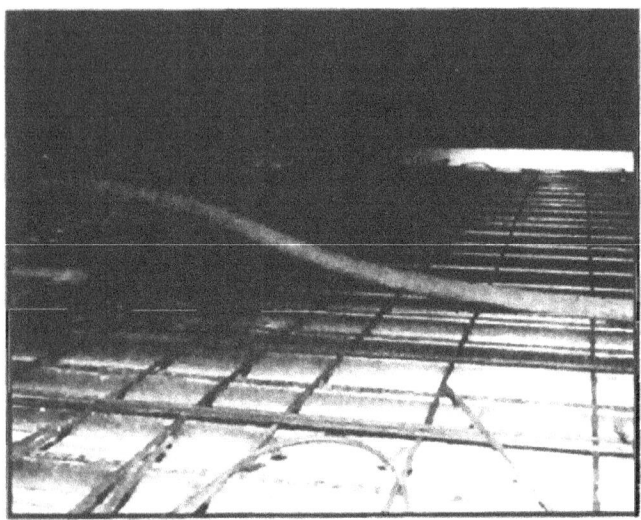

Figure 29.—Top view of overcast deck showing reinforcing bars.

The two wing walls sit on the edges of the top deck parallel to B-drift (figure 25) and seal the area between the top deck and the overhanging roof line. In the LLEM installation, the wing walls seal the void space above the top deck toward A- and C-drifts in crosscut 3, thereby forcing the ventilating air to flow under the overcast deck in the crosscut and above the deck in B-drift. Vinyl straps attached to the top of the vinyl bladders (wing wall grout forms) are tied to 150-mm-long by 25-mm-diam Hilti Dyna bolts anchored in the mine roof. No spreader bars, piers, or chains are used for the top wing walls. The grout is pumped into the inflated vinyl wing wall bladders in the same manner as the side walls, with all of the air and water completely bled from the bladder as the grout fills the bladder. Figure 31 is a side view of the installation of a wing wall. Figure 32 is an end view of a wing wall on top of the overcast deck. The wing walls are then attached to the deck by 900-mm-long by 25-mm-diam full-resin bolts drilled up through the bottom of the overcast top deck and into the bottom of the wing walls. These bolts are spaced on 1.5-m centers. The top edge of the wing walls are then bolted into the mine wall overhang using the same type of bolts on the same spacings.

Two metal W-straps are bolted to the inside of the overcast side walls by 150-mm-long by 25-mm-diam Hilti Dyna bolts and then into the mine ribs using 900-mm-long by 25-mm-diam full-resin bolts. Figure 33 is a view of the completed overcast from underneath the deck. Three steel angle braces were anchored to the outside of each of the two overcast side walls and to the concrete mine floor using the Hilti Dyna bolts. The use of these angle braces is not standard installation practice; they were used in the LLEM evaluations since the side walls could not be directly recessed into the mine ribs because of the larger size of the intersection. A final coating of shotcrete was applied to the entire overcast structure to seal the wing wall structures to the mine strata and to seal all bolted areas.

Instrumentation was then placed on the side walls, deck, and wing walls of the completed overcast. Steel angle frames were erected beside the outby side wall in B-drift to support the instrumentation boxes and LVDTs, as shown in figure 34. One LVDT was placed at the center of the side wall, one at quarter height and midwidth, one at three-quarters height and midwidth, and one at midheight at one-quarter of the way from the side (figures 25B and 34). This was the same positioning used for the LVDTs on the seals. The LVDTs on the deck were suspended from the roof by steel frames (figure 35). A detail of the LVDT attached to the top of the deck is shown in figure 36. A top view of the deck showing LVDT positions is shown in figure 25C. One LVDT was also positioned at the middle of the wing wall facing C-drift.

Figure 31.—Construction of wing wall along edge of overcast deck: side view of vinyl bladder for wing wall being installed.

Figure 32.—Construction of wing wall along edge of overcast deck: end view of wing wall above deck.

Figure 33.—Completed overcast viewed from under the deck.

Figure 34.—Side wall of overcast, as viewed from B-drift outby, showing instrumentation boxes and support frames.

Figure 35.—Instrumentation on top of overcast deck: three LVDTs suspended above deck.

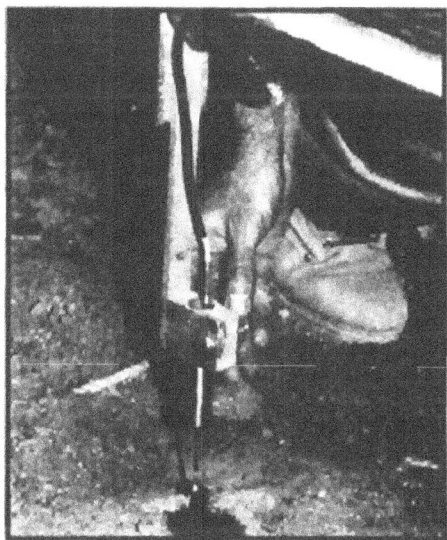

Figure 36.—Closeup of LVDT suspended from roof and attached to deck.

EXPLOSION AND AIR LEAKAGE TEST RESULTS

A summary of the seven explosion tests of the Barclay Mowlem program can be found in table 1, which lists average maximum explosion pressures and flame speeds for each test. Results of the seal, stopping, and overcast evaluations are listed in table 5. More detailed data for the explosion tests are found in the appendices. Tables B-3 through B-9 list the static pressures at the various instrument station locations and the interpolated static pressures at the seals. Also listed are the total pressures at the seals measured by the transducers that were positioned directly in front of the seals. For the weaker explosions (LLEM tests 358 and 361-363), the total pressures at the seals averaged about 15% to 20% higher than the static pressures. For the stronger explosions (LLEM tests 359-360 and 364), the total pressures averaged about 20% higher than the static pressures, but there was more variation in the ratios. The summary of flame arrival times at the various stations for each explosion are in table C-2 in appendix C. These flame arrival times were used to calculate the average flame speeds in table 1. The summary tables of LVDT data are in appendix D. The LLEM explosion tests and the corresponding stopping, seal, and overcast evaluations are discussed in detail in the following paragraphs.

Before the first explosion test, the 450-mm-thick seal in crosscut 2 (figure 17) and the 170-mm-thick water-filled tube stopping (figure 37) in crosscut 3 were evaluated for air leakage resistance. The four differential pressures listed in table A-5 correspond to the four speeds of the main ventilation fan at the LLEM. As can be seen in table A-5, virtually no air leakage could be detected through the crosscut 2 seal for pressure differentials up to 1 kPa. The water-filled tube stopping showed significant air leakage up to 0.55-kPa pressure differential. At 1-kPa pressure differential, the water tubes separated from the bottom support (figure 38). Note that this stopping was designed to release in this manner during low-level explosion pressures. These water-filled tubes were reattached to the stopping base and shotcrete was reapplied. The air leakage through this stopping was then remeasured before the first explosion test, as listed in table A-5. The air-inflated vinyl bladder used for the quickseal was installed in crosscut 4 just before the first explosion test to evaluate its ability to withstand a low-level pressure pulse without any rupture to the vinyl bladder (figure 39).

First Explosion Test (LLEM Test 358)

On February 11, 1998, the ignition of the 8.2-m^3 methane zone at the face of C-drift generated an average pressure pulse of 22 kPa (3.2 psi) during the first Barclay Mowlem explosion test. These pressure values are based on a 10-ms time average (15-point smoothing) of the raw pressure signals and were measured over the length of entry that contained the seal and stopping designs, i.e., from the pressure transducer just inby crosscut 2 to the pressure transducer just outby crosscut 4. The complete listings of the peak static pressures (P_{max}) at the various transducer locations for test 358 are in table B-3. Also listed in the table are the integrals of the pressure over time, $\int Pdt$, for

Figure 38.—Release of individual water tubes of stopping during air leakage test.

Figure 39.—Air-inflated vinyl bladder of quickseal in crosscut 4.

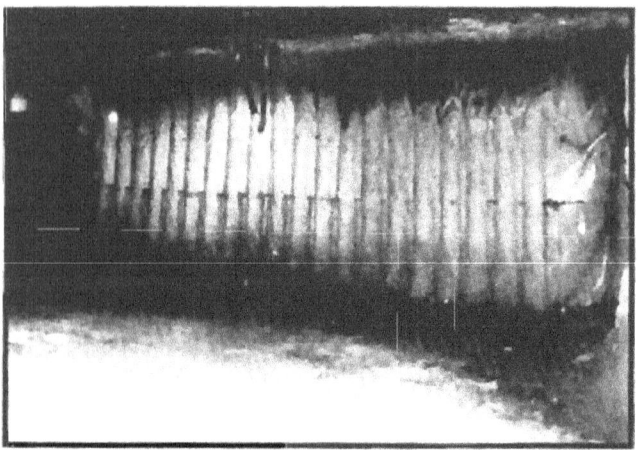

Figure 37.—Completed water stopping in crosscut 3.

each station. Beneath table B-3 are the interpolated static peak pressures at the seal and stopping locations for this test. In addition to the static pressure data interpolated from the transducers mounted into the data-gathering panels at the mine rib, a strain gauge transducer was mounted in front of each seal and stopping design. These transducers approximately measured the total explosion pressure (static plus dynamic), as discussed in the "Instrumentation" section of this report. For the water-filled tube stopping in crosscut 3, this crosscut transducer measured a total pressure of 23 kPa (3.3 psi) compared to the static pressure of 19 kPa (2.8 psi) obtained from an interpolation of the inby and outby transducer data. For the quickseal in crosscut 4, the interpolated static pressure was 14 kPa (2.0 psi). The pressure trace from the total pressure transducer in front of this stopping had too much noise to obtain a reading.

An important measure of the damaging potential of the explosion pressure pulse is the total pressure impulse, which is the time integral of the pressure trace ($\int Pdt$) multiplied by the surface area of the seal. Therefore, the total impulse is $\int PAdt$, where P is pressure, A is the area of the seal, and t is time. The $\int Pdt$ data are listed in table B-3, and the areas of the seals are listed in table 4. The destructive forces of the explosion blast wave depend on both the maximum peak overpressure and the impulse [Sapko et al. 1987]. Under the current U.S. evaluation criterion, a seal design need only withstand a minimum static pressure pulse of 138 kPa (20 psi) while maintaining acceptable air leakage resistance (table 2); impulse requirements have yet to be defined. For this reason, seal designs in previous research programs were frequently subjected to higher level explosion pulses in the LLEM as a means to evaluate the various seal designs against higher impulse loadings. The calculated pressure-time integral for either static or total pressure for the seal in crosscut 2 was about 9 kPa-s, giving a total impulse of 110 kN-s.

Postexplosion observations showed, as expected, no structural damage to the seal in crosscut 2. However, a portion of the shotcrete coating had been removed by the explosion, exposing some of the vinyl bladder. The water-filled tube stopping in crosscut 3 withstood the explosion pressure (figure 40). The individual water tubes released, water was dispersed throughout the area, and the pressure was successfully vented as designed. However, all except two of the water tubes unexpectedly ruptured along each bottom tube seam, resulting in a complete loss of water from each tube. Under ideal performance, about one-half of the water would have remained in each tube after the explosion. The tubes, although emptied of water, performed as expected and restored partial ventilation. The quickseal air-inflated vinyl bladder in crosscut 4, although dislodged, withstood the low-level explosion without rupturing. Because of a program change, the standard anchoring techniques to the mine strata and the remote deployment method were not implemented with the quickseal design as originally planned.

Postexplosion air leakage tests were not performed against the water stopping or the quickseal air-inflated bladder since both were designed to release during the explosion and air loss would be expected. The air leakage through the crosscut 2 seal was well within the established guidelines for this program, as can be seen in table A-6.

After explosion test 358, the two stoppings were removed from crosscuts 3 and 4, and a new seal was built in the high roof area of crosscut 3 (see tables 3-4). This seal (figure 41) was similar to that previously installed in crosscut 2 except that the roof spreader bar was not recessed into the roof. A preexplosion air leakage test was conducted against the newly installed seal in crosscut 3. The results (table A-6) showed that the seal was well within the established guidelines (table 2). In order to more fully evaluate the strengths of these seals and to generate data to assist in developing a numerically based design tool for explosion seals, successive and more intense explosions were required.

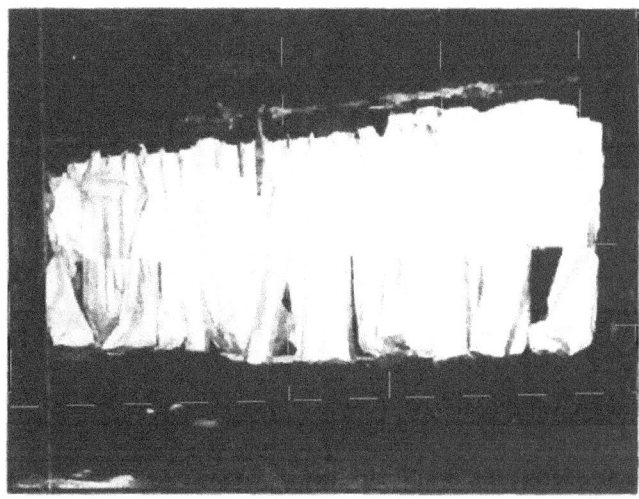

Figure 40.—Condition of water stopping in crosscut 3 after test 358.

Figure 41.—Completed seal in the high roof section of crosscut 3.

Second Explosion Test (LLEM Test 359)

The second explosion test, on February 27, 1998, was designed to produce a static overpressure of at least 140 kPa at the locations of the seals in crosscuts 2 and 3. The seal designs would be rated a "type C" design based on the Queensland Department of Mines and Energy's [1996] Approved Standard for Ventilation Control Devices if the designs could successfully withstand this explosion overpressure. A type C seal design is a typical design used for most circumstances in both Australian and U.S. mines. In order to achieve the desired overpressure of at least 140 kPa, some added pulverized coal dust was loaded onto four shelf locations suspended from the mine roof on 3-m spacings starting from the end of the gas zone.

An average static pressure pulse of 185 kPa (27 psi), as measured over the seal test zone, was generated during this second test (table 1). The flame speed averaged 225 m/s over this zone based on the flame arrival times in table C-2. Figure 42 shows the static pressure traces (10-ms time averaged) at various distances down the entry for test 359. The peak static pressure (table B-4), as interpolated from the inby and outby transducers, was 205 kPa (30 psi) for the crosscut 2 seal and 170 kPa (25 psi) for the crosscut 3 seal. The "total" pressure values measured by the transducers located in the crosscut in front of each seal were 270 kPa (39.5 psi) for the crosscut 2 seal and 220 kPa (32 psi) for the crosscut 3 seal in the high roof area.

Postexplosion observations revealed only minor outward damage to the two seals in terms of hairline cracks and flaking of the shotcrete coatings (figures 43-44). The maximum displacements for all of the LVDTs on the two seals are listed in table D-2 in appendix D. The greatest movement for seal 2 was 10.7 mm for the LVDT at the right-side position (farthest outby) on the back of the seal. The movement at the middle position on this seal was 5.6 mm. The greatest movement for seal 3 was 4.5 mm for the LVDT at the right-side position. The movement at the middle position on this seal was 2.0 mm. Subsequent air leakage tests (table A-7) provided marginal results compared to the established U.S. guidelines. The crosscut 2 seal exceeded the leakage guidelines for all pressure differentials >0.25 kPa, and the crosscut 3 seal in the high roof area exceeded the guidelines for a pressure differential >0.75 kPa. Smoke tube evaluations revealed the location of the main air leaks for both seals. One air leak was at the grout injection port on the crosscut 2 seal; the other leaks were along the riblines behind the steel strapping on both seals. This may not pose a significant problem in an actual coal mine installation since the seals will be recessed into the ribs with subsequent shotcrete coatings.

Third Explosion Test (LLEM Test 360)

On March 3, 1998, a third explosion test was done to subject the two similar seal designs to an explosion overpressure >345 kPa (50 psi). The seals would be rated as "type D" based on the Queensland Department of Mines and Energy's Approved Standard for Ventilation Control Devices if they successfully

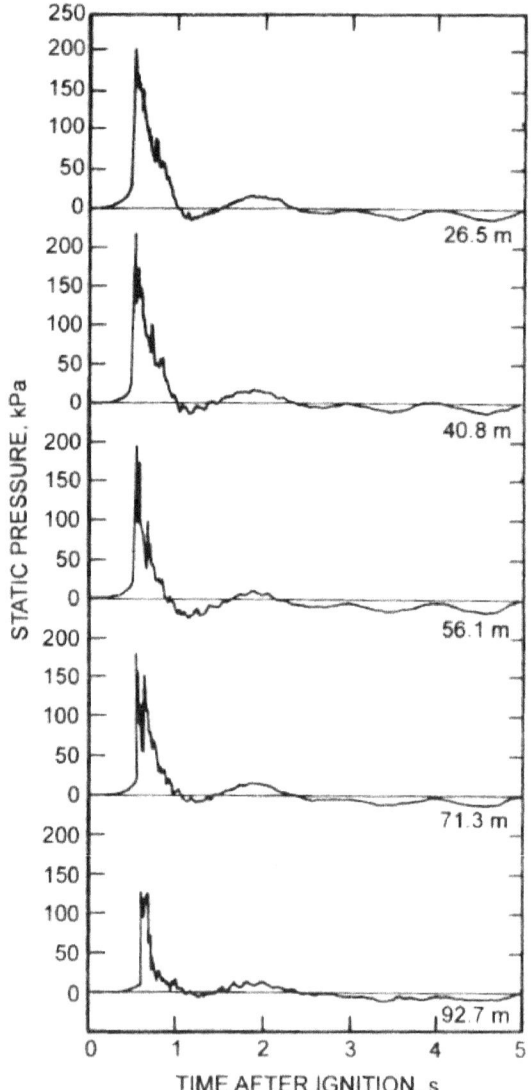

Figure 42.—Pressure traces as a function of distance from the closed end (face) in C-drift for test 359.

Figure 43.—Seal in crosscut 2 after test 359.

Figure 44.—Seal in crosscut 3 after test 359.

Figure 45.—Seal in crosscut 2 after test 360.

withstood this explosion overpressure. Using type D seal designs would allow personnel to remain underground in Australia even if an explosive atmosphere and potential ignition source existed within the sealed area.

As described previously in the "Mine Explosion Tests" section, the setup for this third explosion test was similar to that for the second explosion test except that the coal dust amount was increased from 14.5 to 120 kg and the dusted zone was extended out to 78 m from the face. The average explosion overpressure generated from this third explosion test was 435 kPa (63 psi), this was taken from the transducers in the data-gathering panels located just inby the crosscut 2 seal to just outby the crosscut 3 seal (table 1). The average flame speed for this explosion test was 385 m/s. The explosion generated a 370-kPa (54-psi) static pressure and a 380-kPa (55-psi) "total" pressure against the seal in crosscut 2 (table B-5). However, for the seal in crosscut 3, the peak static pressure was 475 kPa (69 psi) and the "total" pressure was 545 kPa (79 psi). This variation in pressure on the two seals was probably due to pressure piling as the explosion traveled through the coal dust zone. Based on the very limited number of explosion tests in the LLEM that generated peak pressures >250 kPa, no guarantee could be provided during this program on the ability to achieve a uniform ~345-kPa pressure pulse throughout the seal test zone. Future studies may attempt to better achieve this goal through experimentation, such as varying the coal dust loading through the zone.

Postexplosion observations showed that the seal in crosscut 2 survived the 370-kPa static overpressure with only minor damage to the surface shotcrete coatings (figure 45). Subsequent air leakage evaluations for this seal (table A-8) showed that the leakage rates were at or slightly above the established air leakage guidelines. An interesting result of this test was that the postexplosion air leakage for seal 2 (table A-8) was lower than the preexplosion leakage (table A-7). For example, the postexplosion air leakage was 32% lower at 1-kPa pressure differential. One potential explanation for this phenomenon is that if a seal is not significantly damaged during an explosion,

the high-pressure dust-laden gas-air mixture is forced into and through the small orifices and hairline cracks in and around the seal. Some of these small dust particles are then trapped within the cracks and orifices, essentially plugging them. This postexplosion seal plugging was also observed while conducting explosion tests of seals in a chamber as an alternative approach for determining the ultimate strength characteristics of mine seals [Sapko and Weiss 2001]. Figure 46 shows the static pressure data from the two transducers on either side (41- and 56-m positions) of the seal in crosscut 2 compared to the total pressure data from the transducer in front of the seal (48 m). In the figure, the time scales of the two static pressure traces were shifted so that the peaks matched with that of the transducer at the seal. In the upper part of figure 46, the traces from the LVDTs at the middle and bottom positions on the seal are shown. They show the seal being displaced by the pressure pulse and then returning to its original position as the pressure decays. The maximum displacements for all of the LVDTs on this seal are listed in table D-3. The greatest seal movement was 12.2 mm for the LVDT at the right-side position (farthest outby) on the back of the seal. The movement at the middle position on the seal was 5.6 mm.

The seal in the high roof area of crosscut 3 was destroyed by the 475-kPa (69-psi) static pressure pulse (figure 47). From postexplosion observation of the seal, it seemed that the seal failure started at the top of the seal, where only a minimum of hitching was achieved due to the contour of the hard limestone mine roof. Figure 48 shows the static pressure data from the two transducers on either side (71 and 93 m) of the seal in crosscut 3 compared to the total pressure data from the transducer in front of the seal (75 m). As in figure 46, the time scales of the two static pressure traces were shifted so that the peaks matched with that of the transducer at the seal. In the upper part of figure 48 are the traces from the LVDTs at the middle and bottom positions on the seal. The maximum displacements for all of the LVDTs on this seal up to the time of seal failure are listed in table D-3. The greatest seal movement was ~30 mm at three of the LVDTs. The times of seal failure

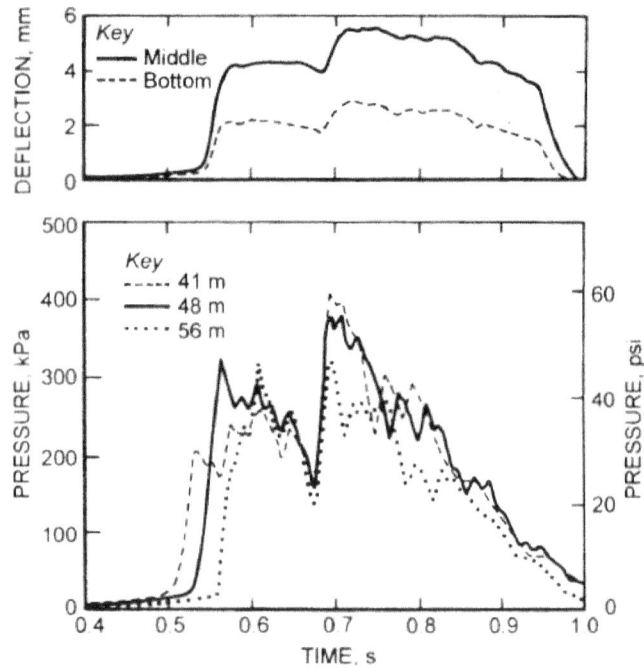

Figure 46.—Pressure and LVDT traces for seal 2 during test 360.

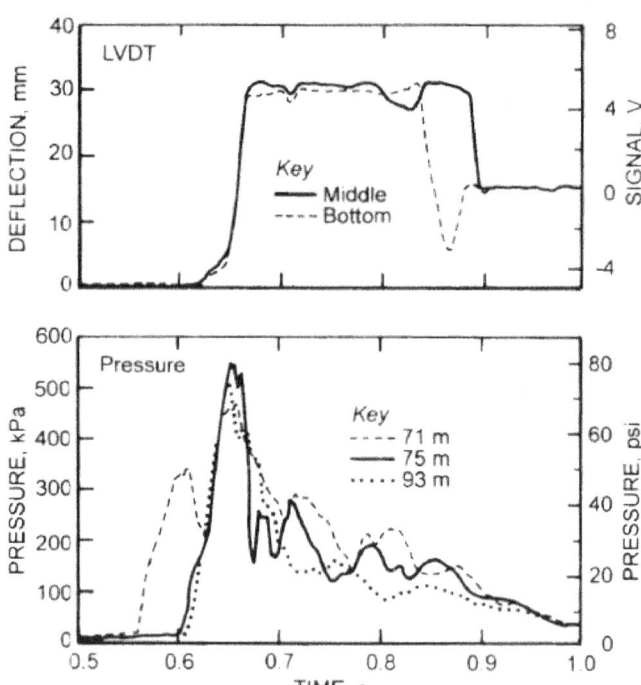

Figure 48.—Pressure and LVDT traces for seal 3 during LLEM test 360.

Figure 47.—Remains of crosscut 3 seal after test 360.

can be determined from the LVDT traces. For LLEM test 360, the LVDTs were set at an initial position corresponding to about -5 V. This would allow them to measure a large signal up to about +10 V, or a total movement of about 45 mm. The sudden drop in the LVDT signals and return to ~0 V (~15 mm) in figure 48 signifies that the seal failed and the LVDTs were destroyed. Failure occurred at ~0.84 s for the bottom LVDT and at ~0.89 s for the middle LVDT. The LVDT traces for the seal in crosscut 3 do not gradually return to their original positions as they did for the seal in crosscut 2 (figure 46). The rubble from the destroyed seal in crosscut 3 was removed before the next series of tests against the overcast.

Fourth, Fifth, and Sixth Explosion Tests (LLEM Tests 361, 362, and 363)

The next three low-level explosion tests (see tables 1 and 5) were designed to evaluate the overcast constructed at the intersection of crosscut 3 and B-drift during March 7-24, 1998. The explosions originated at the face of C-drift, and the pressures traveled down C-drift and then through crosscut 3 to the underside of the overcast (figure 25). The fourth explosion test (LLEM test 361) on March 26, the fifth test (LLEM test 362) on March 31, and the sixth test (LLEM test 363) on April 1, 1998, were designed to generate overpressures ranging from about 15 kPa to 50 kPa at the overcast location. This is the first time that an overcast design has ever been explosion tested under full-scale conditions. Before conducting the first explosion test against the overcast design, an air leakage test (table A-8) was done. To accomplish the air leakage test on the overcast, both B- and C-drifts were sealed outby the overcast location using brattice curtain attached to a wooden framework custom-fit to the entry. In addition, the crosscut between A- and B-drifts in crosscut 3 was sealed, as was B-drift inby the overcast location. The pressure differential under (or through) the overcast increased as the ventilation fan speed increased. Any air leakage through the overcast deck, side walls, or wing walls was detected through the 465-cm^2 center opening in the B-drift inby curtain. The air leakage rates through the overcast design (table A-8) were well within the air leakage guidelines established for seal designs. The overcast withstood the 16-, 30-, and 47-kPa explosion overpressures generated during tests 361, 362, and 363, respectively (see tables 1, 5, and

B-6 to B-8). These pressures were measured by a transducer located in the inby rib of crosscut 3. Therefore, they are the static pressures that would be exerted on the inner sides of the side walls and the underside of the deck of the overcast. Cracking of the cementitious top deck became more pronounced after each of the three explosion tests and was a result of the upward pressure exerted on the deck as the explosion pressure vented through the overcast. However, the leakage rates (table A-9) through the overcast after the 16-kPa (2.3-psi) explosion test 361 were still well within the established guidelines. After this leakage test, shotcrete was reapplied to the overcast on all visible cracks. These cracks were mainly on the overcast deck and the interface of the deck with the side walls and wing walls. The leakage rates from a subsequent air leakage test showed only nominal improvements compared to those of the first test. No further air leakage tests were done for the overcast design after LLEM test 361. The LVDT data for the overcast tests are listed in tables D-4 through D-6. The data show almost no movement (<1 mm) of the side wall in any of the tests. The LVDT on the wing wall recorded a maximum movement of 2.6 mm during LLEM test 363. The maximum upward movements of the deck in LLEM tests 361, 362, and 363 were 5.5, 14.5, and 16.4 mm, respectively. The LVDTs located off-center on the deck usually recorded smaller movements than the middle LVDT. The upward movements of the deck were followed by maximum downward rebound movements of 1.4, 15.7, and 15.7 mm in LLEM tests 361, 362, and 363, respectively. Examples of the LVDT traces for the overcast are compared to the pressure trace for LLEM test 363 in figure 49. The static pressure was measured at the inby crosscut rib leading from C-drift to the underside of the overcast. The LVDT traces include the one on the wing wall, two on the deck, and the middle one on the side wall (see table D-6 for summary data and figure 25 for locations). In figure 49, positive deflection on the wing wall is toward C-drift. On the deck, positive deflection is upward; on the side wall, positive deflection is outby.

Seventh Explosion Test (LLEM Test 364)

The seventh and final explosion test of this program was done on April 3, 1998, to evaluate the capability of two additional seal designs to withstand a pressure pulse of at least 140 kPa. This test also provided additional evaluation of the overcast. The test setup was similar to that of the second explosion test (LLEM test 359) in terms of gas ignition zone and coal dust loading on each shelf, except that only two shelves were used instead of four. The resultant explosion generated an explosion overpressure that averaged 160 kPa (23 psi) through the seal test area. Detailed data on the pressures at various distances down the entry and at the seals are listed in the table B-9. The flame speed averaged 340 m/s for this test.

The 240-mm-thick seal in crosscut 3 consisted of individual bladder tubes attached to a single vinyl unit. This unit was suspended using chains from the steel framework bolted to the mine roof. This chain was also installed into each of the bladder

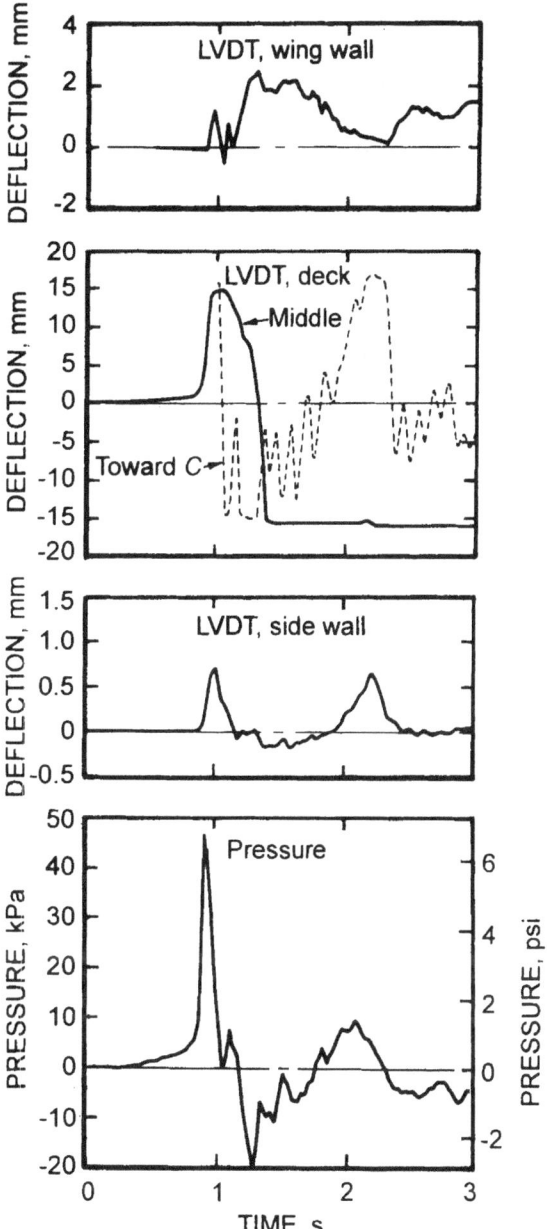

Figure 49.—Pressure and LVDT traces for overcast during LLEM test 363.

tubes before the grout injection into the tubes. The back side of the completed seal is shown in figure 50, with one LVDT in the center of the seal. The seal in crosscut 4 (figure 51) was very similar in design, but used 165-mm-diam tubes. This smaller seal design was installed in crosscut 4 before the low-level explosion tests against the overcast design to allow for a longer cure period for the grout. The new 240-mm-thick seal in the high roof section of crosscut 3 was installed after the overcast evaluation tests. This seal was explosion tested about 24 hr after its completion.

Before this seventh and final explosion test of the Barclay Mowlem program, shotcrete was reapplied to seal 2. The seals

Figure 50.—Completed new (second) seal in high roof section of crosscut 3 (back side).

Figure 52.—Remains of crosscut 4 seal after test 364.

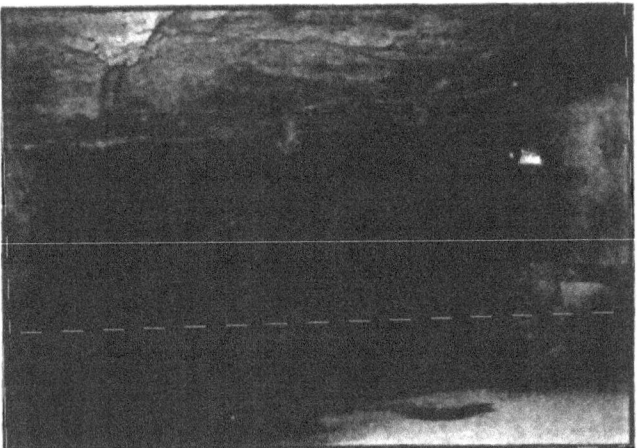

Figure 51.—Completed new seal in crosscut 4.

were then leakage tested (table A-10). No air leakage could be detected through any of the three seals at pressure differentials up to 0.34 kPa. For pressure differentials of 0.55 and 1.06 kPa, only minor leakages were measured, but they fell well within the established guidelines for these seal evaluation programs. The data in table A-10 for seal 2 show that the additional shotcrete significantly reduced the air leakage from the values measured earlier (see table A-8).

The ignition of the gas zone and subsequent burning of the coal dust generated pressures of 195 kPa (28 psi) at the 450-mm-thick seal in crosscut 2, 160 kPa (23 psi) at the 240-mm-thick seal in the high roof area of crosscut 3, and 115 kPa (17 psi) at the 165-mm-thick seal in crosscut 4. Both new seals in crosscuts 3 and 4 were destroyed by the explosion. As figure 52 shows, the individual grout columns of the 165-mm-thick seal in crosscut 4 were apparently not thick enough to withstand the explosion pressure pulse, which resulted in nearly complete destruction. Except for a few of the grout-filled tubes near each rib, most of the grout-filled tubes separated from the floor anchor and sheared at a point just below the top reinforcement on each tube. Nearly all of the shotcrete material filling the top section of the seal above the grout-filled tubes was displaced by the explosion. There were similar findings for the 240-mm-thick seal in crosscut 3. Both of these new seals were significantly thinner than the 450-mm-thick seals, which had been previously tested and survived a similar explosion (tables 4 and 5). In addition, the new seals did not have the additional reinforcement in terms of the concrete piers and metal strapping that was incorporated into the first two seals. The LVDTs on the seals measured maximum displacements of 12.6 and 17.0 mm on seals 3 and 4, respectively, before seal destruction (table D-9). The 450-mm-thick seal in crosscut 2 survived this explosion as it had the previous ones. As shown in table A-11, the postexplosion air leakage rates for this seal were well within the acceptable guidelines.

It was originally anticipated by Barclay Mowlem that the seal in crosscut 3 and probably the seal in crosscut 4 would survive this explosion, and therefore the pressure pulse would continue outby in C-drift until it reached the next open crosscut (crosscut 5). It would then vent through the crosscut and travel back inby in B-drift. The pressure pulse would then pass over the overcast (figure 25) and exert a downward pressure load on the overcast deck. Therefore, before the test, the LVDTs were moved from positions on the outby face of the side wall and the top of the deck to positions under the overcast. One LVDT was positioned underneath the deck in the middle of the deck, and a second LVDT was positioned halfway between the middle and the C-drift edge of the deck. No LVDTs were placed on the side wall. As a result of the failure of the seal in crosscut 3, the pressure pulse traveled down crosscut 3 and the overcast was subjected to an overpressure of 41 kPa (6 psi) under the overcast deck. The maximum upward movement of the overcast deck during this test was 15.8 mm; the details on the other measurements are in table D-8. This upward deflection was comparable to that observed in LLEM test 363, which had a slightly higher pressure underneath the overcast. Because the pressure pulse came under the deck rather than above as originally expected, the two LVDTs were torn from their supports during the test. However, the middle LVDT on the deck was able to record both the initial upward deflection and the later downward rebound before being destroyed. The LVDT toward C-drift may have been destroyed before recording the downward rebound. The overcast itself survived explosion test 364, as it had test 363.

PRELOADED SOLID-CONCRETE-BLOCK SEAL DESIGNS FOR FRIABLE RIB CONDITIONS

In 1998, PRL and MSHA jointly participated in a research program to evaluate the strength characteristics and air leakage resistance of preloaded solid-concrete-block seals for use in underground coal mines. These seals were specifically designed for use in areas of mines where the standard method of hitching or keying of the concrete block seal into the mine ribs was impractical due to the weak coal (i.e., friable ribs). These seals were jointly designed by personnel from the United Mine Workers of America (UMWA) at Jim Walter Resources, Inc.'s Blue Creek No. 5 Mine; Strata Products, Inc.; and Jim Walter Resources, Inc., in Alabama. Assistance was provided by MSHA and PRL personnel.

Installation of the standard solid-block seal requires floor and rib hitching to meet the intended explosion pressure resistance of 138 kPa (20 psi). Standard seal strength is due to an arching action that occurs through the thickness of the seal, which applies lateral thrust to the coal ribs, floor, and roof. Jim Walter Resources, Inc.'s No. 5 Mine encountered extremely soft and friable ribs conditions, making conventional rib hitching of the standard seal almost impossible. The coal was so soft that large sections of the rib coal can be removed by hand, making it very difficult to interface with competent coal. Injection grouting to strengthen the ribs was not considered economically feasible.

The alternative seal designs tested under this program were based on better coupling between the seal and the mine roof and ribs by preloading the seal with pressurized grout bags. These pliable fiberglass Packsetter bags were placed along the perimeter of the seal and pressurized with a cementitious grout.

CONSTRUCTION

Three solid-concrete-block seals were installed in the LLEM (tables 3 and 4). The seals were very similar to the standard-type solid-concrete-block design described in the CFR. Each seal used solid concrete block with nominal size of 20 cm by 20 cm by 40 cm (8 in by 8 in by 16 in). These were tongue-and-groove solid blocks supplied by Willcut Block & Supply Co., Inc., of Tuscaloosa, AL. Based on a random sampling, the average weight per block was 26.4 kg. Two of the three seals used fully mortared joints (seals in crosscuts 2 and 3); one seal used only dry-stacked blocks (seal in crosscut 4). The Quikrete portland cement mortar was from Independent Cement Corp., Hagerstown, MD. This mortar exceeds the compressive strength requirements of ASTM C-387 and ASTM C270, type N. For each seal, the vertical block joints between courses were staggered, the main seal wall was 405 mm thick, and an interlocked 81-cm by 40-cm center pilaster was used. The seals did not use any rib hitching. Floor hitching was used to augment the strength of one seal (crosscut 2); no floor hitching was used for the other two seals. The floor hitching for the seal in crosscut 2 was simulated by anchoring 15-cm by 15-cm by 13-mm-thick steel angle to the mine's concrete floor in a position abutting the lower course of concrete block. The steel angle was anchored to the floor using 25-mm-diam by 230-mm-long Hilti Kwik Bolts II on 46-cm centers. Any gaps between the steel angle and the seal were filled with mortar. The mortared block seals in crosscuts 2 and 3 required 276 and 300 concrete blocks, respectively. The dry-stacked design in crosscut 4 required 355 blocks. For the seals in crosscuts 2 and 3, half-size concrete block (10-cm by 20-cm by 40-cm) were used to minimize the gap near the mine roof.

The main difference between these concrete block seals and the design described in the CFR is the use of pliable bags pressurized with cementitious grout on the mine ribs and roof in place of conventional hitching. These pressurized bags are referred to as "Packsetter bags." The Packsetter grout bags, manufactured by Strata Products, Inc., Marietta, GA, have a 130-cm by 80-cm outer shell of plastic weave with a 122-cm by 76-cm inner plastic bladder. The Packsetter bags have a one-way valve constructed in the filling port. This valves permits the flow of grout into the bag and closes to prevent the grout from flowing back out of the bag. The Packsetter bags are positioned at the interface between the seal and the roof and ribs (figure 53) and then filled with grout (figure 54). For the seals in crosscuts 2 and 3, 11 full-size and 1 half-size Packsetter bag (figure 54) were required along the seals' interfaces to the mine roof and ribs. For the dry-stacked seal in crosscut 4, 10 full-size and 10 half-size Packsetter bags were used. For this seal, the interface with the mine roof used a combination of full- and half-sized bags. The bags were overlapped a minimum of 15 cm. The distance between the mine rib and block should be <2-5 cm. During the LLEM evaluations, the blocks were installed tight against the bag and rib (figure 55). The gaps between the top block course and mine roof ranged from 5 to 12 cm.

To facilitate the construction process, the grout was injected into the Packsetter bags using the mine's compressed air to power the grout pump. As an alternative method for filling the bags where a compressed air supply may not be available, a few of the Packsetter bags during this construction process were

Figure 53.—Unfilled Packsetter bags at the seal interface with the mine roof and ribs.

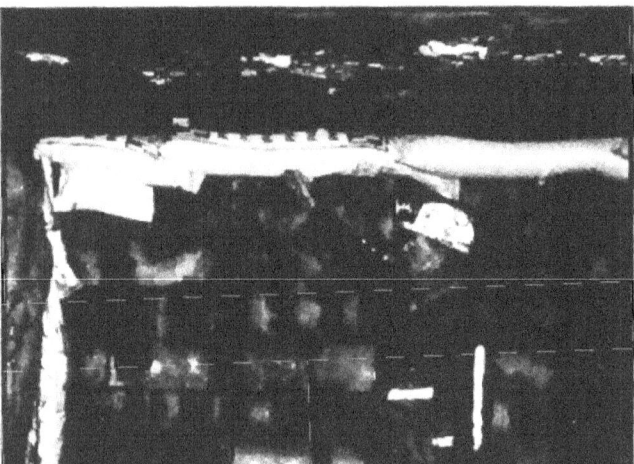

Figure 54.—Filled and pressurized Packsetter bags at the outby roof and rib seal interface showing full-size bags and one half-size bag on the left.

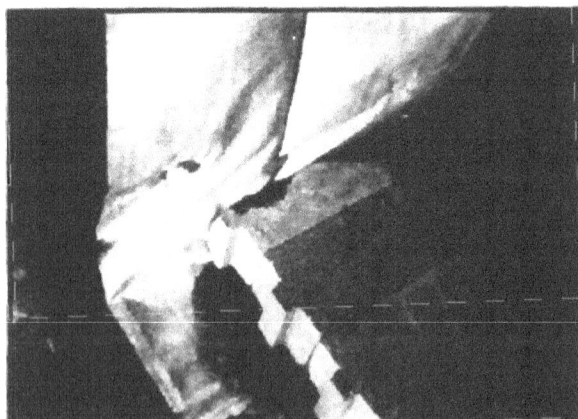

Figure 55.—Placement of the Packsetter bag at the mine rib and floor interface with the bottom course of the tongue-and-groove solid-concrete-block seal.

filled with grout using a hand pump unit (figure 56). The Packsetter grout is a specially formulated portland cement-based mixture that is blended and packaged for Strata Products, Inc., by Quikrete in Virginia. One of the key components of the grout is calcium aluminate, which decreases curing times and increases the compressive strengths compared with conventional portland cements. The compressive strength of the Packsetter grout is 2.5 MPa (362 psi) after 24 hr, 3.0 MPa (435 psi) after 7 days, and 4.0 MPa (580 psi) after 28 days. This grout is a high-yield grout that requires significant amounts of water compared to conventional cements. About 55 L of water is required per 23-kg bag of Packsetter grout. The Packsetter bag is designed to contain the entire amount of water with no seepage to meet the maximum specification of 2% free water after mixing with grout is complete. The grout is also classified as a nonshrink grout, which specifies <1% shrinkage during the cure period. This is a critical specification required when using the grout in a prestressing operation. The Packsetter bags were filled with grout to an internal pressure of 250-275 kPa (36-40 psi) for the seal in crosscut 3 and ~300 kPa (~44 psi) for the seals in crosscuts 2 and 4. The Packsetter bags along the mine roof were injected first (starting at the center and working toward the ribs), followed by the rib bag closest to the mine floor on each side of the seal. The remaining rib bags were then filled in no particular order. When injected with grout, the Packsetter bags overlapped both sides of the block wall a minimum of 8 cm (figure 54). Approximately 8-10 bags of grout were used to fill the Packsetter bags for each the seals in crosscuts 2 and 3. The dry-stacked seal in crosscut 4 required 16 bags of grout.

The completed Packsetter seals in crosscuts 2 and 3 are shown in figures 57-58. Upon completion of the Packsetter seals, sealant was applied to selected perimeter areas on both sides of the seals in crosscuts 2 and 3. The sealant was applied

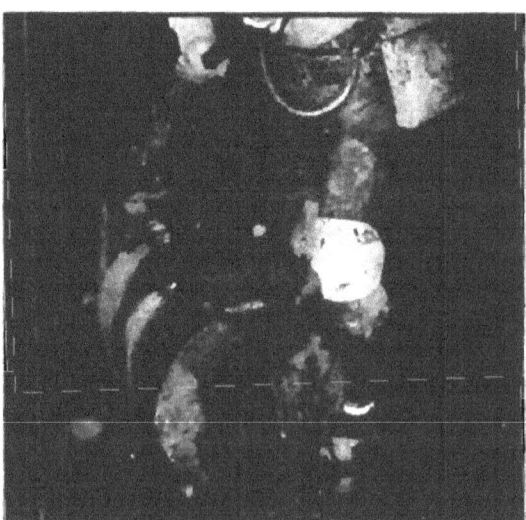

Figure 56.—Hand-powered pump for filling the Packsetter bags.

to the block joints and perimeter on both sides of the dry-stacked seal in crosscut 4. The sealant, Quikrete's B-bond, is a fiber-reinforced, surface-bonding cement (MSHA acceptance No. IC-36), which is considered an accepted sealant material [Sawyer 1992]. The LLEM temperatures ranged from 15 °C to 17 °C (60 °F to 62 °F); the relative humidity ranged from 61% to 92% during the seal construction period. The seals had a minimum 30-day cure period before conducting the air leakage evaluations and the explosion tests.

EXPLOSION AND AIR LEAKAGE TEST RESULTS

Before the first explosion test, air leakage evaluations were done on each of the three seals. Each seal exhibited air leakage rates (table A-12) that were well in excess of the rates established for these programs. To reduce the air leakage rates, a full-face coating of sealant was applied to both sides of all three seals. A subsequent preexplosion air leakage evaluation

Figure 57.—Completed mortared seal with the Packsetter bags and floor hitching in crosscut 2.

Figure 58.—Completed mortared seal with the Packsetter bags in crosscut 3.

(table A-13) revealed that the full-face coating of sealant reduced the air leakage rates to a level well within the acceptable maximum limits.

The first explosion test (No. 365) was designed to provide an average static pressure pulse of ~140 kPa throughout the seal test zone (table 1). This was done by igniting a nearly 200-m³ zone of 9.5% methane-in-air at the closed end of C-drift. In an effort to maintain the peak static pressure at the seal in crosscut 1 as close as possible to the 138-kPa requirement of the CFR, coal dust was not suspended from shelving outby the gas zone during this test. Postexplosion observations of the seals after LLEM test 365 revealed no evidence of outward damage to the mortared designs in crosscuts 2 and 3 after being subjected to static pressure pulses of 150 kPa (22 psi) and 130 kPa (19 psi), respectively (see tables 5 and B-10). As with the previous seal evaluation programs, these pressures were rounded to the nearest 5 kPa and 1 psi. The dry-stacked block seal in crosscut 4 was subjected to a 120-kPa (18-psi) static pressure pulse. This dry-stacked seal exhibited a large horizontal crack across the bottom portion of the seal on both sides between the first and second course of block. The block at this level displaced about 13 mm toward B-drift. Subsequent air leakage measurements (table A-14) on these seals showed that the seals in crosscuts 2 and 3 were well within the established guidelines, but the damaged dry-stacked seal in crosscut 4 greatly exceeded the established air leakage rates.

Based on the positive results of the Packsetter bags when used in conjunction with the mortared seals, a second explosion test was done to ensure that the seal in crosscut 3, which saw only 130 kPa (19 psi) during the first explosion test, was subjected to the required 138-kPa static pressure level. To increase the static pressure levels during this second test, 14.5 kg of pulverized coal dust was loaded onto shelving that was suspended from the mine roof just outby the gas zone from 13 to 23 m from the face. The nominal coal concentration of this ~12-m-long dusted zone was ~100 g/m³. The resultant explosion generated a static pressure pulse of 185 kPa (27 psi) at the crosscut 2 seal, 155 kPa (22 psi) at the crosscut 3 seal, and 125 kPa (18 psi) at the crosscut 4 seal (see tables 5 and B-11). The mortared block seal with the Packsetter bags and floor hitching in crosscut 2 survived the explosion with no evidence of outward damage (figure 59). The mortared block design with the Packsetter bags but without the floor hitching in crosscut 3 showed a slight movement at the mine floor, as evidenced by the separation of the sealant applied to the block/floor interface (figure 60). Three vertical hairline cracks were evident on the B-drift (nonexplosion) side of the seal between the center pilaster and the outby rib. Portions of the perimeter sealant on each side of the seal were also dislodged during the explosion. The dry-stacked block design with the Packsetter bags was destroyed by the explosion. Only the bottom course of block, which was mortared to the mine floor for leveling purposes, and parts of the seal near the mine ribs remained (figure 61).

Postexplosion air leakage measurements (table A-15) showed that the mortared block design with the Packsetter bags and floor hitching in crosscut 2 maintained minimal leakages, well within the acceptable rates established in the guidelines for these programs. The mortared block seal with the Packsetter bags, but without floor hitching (crosscut 3), also exhibited air leakage rates within the accepted guidelines. However, before this air leakage test, a small (unauthorized) amount of additional sealant was inadvertently applied by the vendor to the seal at the perimeter areas where the original sealant was dislodged and along the block/floor interface where the sealant had separated.

Based on the positive results (table 5) achieved with two solid-concrete-block seals with full mortar joints, a center pilaster, and the Packsetter bags used in place of the roof and rib hitching, MSHA has deemed these seals to be suitable for use in underground coal mines, especially in areas with friable coal.

Figure 60.—Mortared seal with Packsetter bags in crosscut 3 after test 366.

Figure 59.—Mortared seal with floor hitching and Packsetter bags in crosscut 2 after test 366.

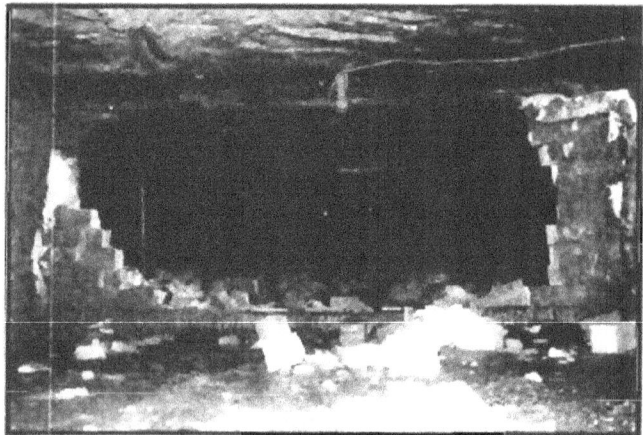

Figure 61.—Remains of the dry-stacked seal with the Packsetter bags in crosscut 4 after test 366.

CONCLUSIONS

As alternative ventilation structure designs and/or construction materials are introduced to the mining industry to address the wide range of geologic, geometric, and environmental conditions encountered in underground coal mines, evaluations of these seal designs and materials must be done to ensure that the designs perform the intended function and that they provide the required protection to underground personnel, as described in 30 CFR 75.335, prior to use in mines. Testing the strength characteristics of these ventilation structures against methane ignitions and/or coal dust explosions and measuring the air leakages in the full-scale LLEM is one accepted method. This report describes several unique and innovative seal, stopping, and overcast designs that were evaluated in three programs during 1997-98.

In the first program, several pumpable cementitious plug seals that do not require floor or rib hitching were subjected to a minimum static pressure pulse of 138 kPa (20 psi) and were shown to meet or exceed the strength requirements mandated by the CFR while maintaining acceptable air leakage rates. These designs were developed by HerTech, a U.S. seal manufacturer that funded the program. These included a 610-mm-thick seal using a pumpable cementitious grout with an average compressive strength of 4.7 MPa, a 760-mm-thick seal with an average grout compressive strength of 3.3 MPa, and a 915-mm-thick seal with an average grout compressive strength of 3.0 MPa. Before this program, any pumpable cementitious plug seal used in an underground coal mine needed to be at least 1.2 m thick with a minimum compressive strength of 1.4 MPa. It was also shown, by the removal of the form wall before the second explosion test in this program, that the form walls used to contain the cementitious grout slurry need not be considered as part of the seal design. However, if these form walls are removed, the exposed cementitious grout must be coated with an approved MSHA sealant.

In the second program, four reinforced cementitious seals and two stopping designs (tables 4 and 5) developed by Barclay

Mowlem Construction Ltd. of Australia were evaluated for strength characteristics and air leakage resistance. This program was funded by Barclay Mowlem. These full-scale designs were air leakage tested, then subjected to a series of explosions. The main objective of the tests was to determine if the seal and stopping designs were of sufficient strength and leakage resistance to meet or exceed the requirements of the Queensland Department of Mines and Energy's Approved Standard for Ventilation Control Devices. This objective was achieved during this program.

A summary of the evaluations is listed in table 5. The two stopping designs withstood the first explosion test. The static pressure exerted on the water tube stopping was 19 kPa; the pressure on the air-inflated vinyl bladder Quickseal, 14 kPa. The vinyl water tubes, although still suspended from the mine roof, had drained of water because of a rupture of the bottom seal of each tube. The Quickseal stopping was dislodged from its original position, but the vinyl bladder was still inflated. During the second explosion test, the static pressures on the two 450-mm-thick seals ranged from 170 to 205 kPa for the seals in crosscuts 3 and 2, respectively. These two seals physically survived the explosion, but postexplosion air leakage resistance data for each seal were near or slightly above the upper limit guidelines established for this program.

A higher level explosion test was then conducted. The 450-mm-thick seal in crosscut 2 withstood a peak explosion pressure of 370 kPa. As had been observed after the previous explosion test, the air leakage resistance data for this seal after this larger explosion test were near or slightly above the upper limit guidelines established for this program. The 450-mm-thick, 2.8-m-high vinyl bladder seal in crosscut 3 was destroyed by a peak pressure of 475 kPa.

A large part of the Barclay Mowlem program was dedicated to evaluating the strength and air leakage resistance of an innovative, full-scale overcast design when subjected to low-level explosion tests. This overcast design withstood four explosion tests, which generated overpressures at the overcast location ranging from 16 to 47 kPa. An air leakage test after the 16-kPa explosion test revealed that the leakage through the overcast design fell well within the established guidelines for air leakage through a seal.

A late modification to the Barclay Mowlem program was the evaluation of a seal design that might be capable of withstanding a 140-kPa explosion overpressure within ~24 hr after construction. Therefore, a second 240-mm-thick seal, composed of individual vinyl tubes, was built in the 2.8-m-high crosscut 3 and explosion tested ~24 hr after construction. This particular seal was destroyed by the 160-kPa explosion pressure generated during the last test of the program. A similar seal in crosscut 4, composed of a series of connected 165-mm-diam vinyl tubes, was also destroyed by this explosion test, which generated an overpressure of 115 kPa at the seal location.

At the request of MSHA and with support from the UMWA and Jim Walter Resources, Inc., an innovative modification to the 405-mm-thick standard-type, solid-concrete-block seal described in the CFR was evaluated in a third program for potential use in areas with friable coal where other seal designs had failed because of the weakness of the rib coal. This modification involved the use of pressurized grout-filled bags, referred to as "Packsetter bags," along the interface of seal with the mine roof and ribs. The purpose of the Packsetter bags was to eliminate the need for hitching of the seal into the mine floor and/or ribs. The two solid-concrete-block designs using full mortar joints, staggered vertical block joints, a center pilaster, and Packsetter bags in place of the rib and/or floor hitching survived the minimum 138-kPa static pressure pulse while maintaining air leakage rates within the acceptable guidelines. A second similar seal design without the mortared block joints (i.e., dry-stacked) was destroyed by a 125-kPa explosion. This mortared Packsetter seal, when used in an entry with friable coal, has been demonstrated at the Jim Walter Resources, Inc., mine in Alabama to provide a seal that can continue its intended function for a longer period compared with other conventional and alternative sealing methods. The mortared solid-concrete-block seals using the Packsetter bags in place of floor and rib hitching have been deemed suitable by MSHA for use under certain conditions in underground coal mines.

NIOSH will continue to develop and/or evaluate, through programs similar to those discussed in this report, new and innovative seal designs that will provide increased protection for U.S. miners. These new seal designs may reduce materials handling, thereby reducing personnel injuries; reduce overall seal installation times, resulting in reduced mine personnel exposure when installing seals under hazardous conditions; and/or enhance seal performance in terms of strength characteristics, air leakage resistance, and better durability in areas of high convergence or unusual geological conditions.

ACKNOWLEDGMENTS

The authors thank John Breedlove, Vice President, Joe Bower, Northern Region Manager, and Rodney Howery, Sales Manager, of HeiTech Corp.; Duncan Hird, Technical Sales Manager, Blue Circle Special Cements, Cedar Bluff, VA; and Phil Smith, Mining Consultant, Blue Circle Industries PLC, Barnstone, U.K. These individuals significantly contributed to the coordination, construction, and evaluation of the cementitious pumpable seals.

The authors also thank Cliff Robinson, Development Manager of Mine Services, Barclay Mowlem Construction Ltd., Moorooka, Queensland, Australia; Barry Sturgeon, Joanne Sturgeon, Collin Furbank, and Leslie Retschlag, formerly with

Barclay Mowlem; and Michael Downs, BHP Coal of Australia. These individuals significantly contributed to the coordination, construction, and evaluation of the high-strength cementitious pumpable seals, overcast designs, and stopping designs.

In addition, the authors thank Joe Main, Administrator, Occupational Health and Safety, UMWA; Kenneth Randall Clements, President, Local 2368 UMWA District 20 during the test program and now Safety Committee Member, Local 2368 of Jim Walter Resources, Inc., Blue Creek Coal No. 5 Mine, Brookwood, AL; Kenneth Howard, Director of Technical Support, and Clete Stephan, Principal Mining Engineer, of MSHA; Thomas McNider, Manager of Ventilation, and Scott Jinks, Mine Engineer, of Jim Walter Resources, Inc., Blue Creek Coal, Brookwood, AL; and Ed Barbour, Sales and Marketing Manager, and Joe Atkins, Application Technician, of Strata Products, Inc., Marietta, GA. These individuals contributed to the coordination, construction, and evaluation of the Packsetter seals.

The authors acknowledge the following PRL personnel without whose contributions this program could not have been accomplished: Kenneth W. Jackson, Electronics Technician, and Deepak Kohli, Electrical Engineer, for sensor calibrations and installations, for modifications to and installation of the data acquisition systems, and for their participation in the testing and data analyses; and William A. Slivensky, Frank A. Karnack, and Donald D. Sellers, Physical Science Technicians, for their extensive participation in the installation of sensor mounting equipment, seal installations, construction monitoring, explosion and air leakage testing, and cleanup.

REFERENCES

CFR. *Code of Federal regulations*. Washington, DC: U.S. Government Printing Office, Office of the Federal Register.

Greninger NB, Weiss ES, Luzik SJ, Stephan CR [1991]. Evaluation of solid-block and cementitious foam seals. Pittsburgh, PA: U. S. Department of the Interior, Bureau of Mines, RI 9382.

Mattes RH, Bacho A, Wade LV [1983]. Lake Lynn Laboratory: construction, physical description, and capability. Pittsburgh, PA: U.S. Department of the Interior, Bureau of Mines, IC 8911.

Mitchell DW [1971]. Explosion-proof bulkheads: present practices. Pittsburgh, PA: U.S. Department of the Interior, Bureau of Mines, RI 7581.

Queensland Department of Mines and Energy [1996]. Approved Standard for Ventilation Control Devices Including Seals and Surface Airlocks, ODM967396. Queensland Department of Mines and Energy, Safety and Health Division, Coal Operations Branch.

Roxborough FF [1997]. Anatomy of a disaster: the explosion at Moura No. 2 coal mine, Australia. Mining Technol 79(906):37-43.

Sapko MJ, Weiss ES [2001]. Evaluation of new methods and facilities to test explosion-resistant seals. In: Proceedings of the 29th International Conference of Safety in Mines Research Institutes. Vol. 1. Katowice, Poland: Central Mining Institute, pp. 157-166.

Sapko MJ, Weiss ES, Watson RW [1987]. Size scaling of gas explosions: Bruceton Experimental Mine versus the Lake Lynn Mine. Pittsburgh, PA: U.S. Department of the Interior, Bureau of Mines, RI 9136.

Sawyer SG [1992]. Mortar for use in the construction of concrete-block stoppings and seals in underground mines. Pittsburgh, PA: U.S. Department of Labor, Mine Safety and Health Administration, Industrial Safety Division (ISD) report No. 02-174-92, June 22, 1992.

Stephan CR [1990a]. Construction of seals in underground coal mines. Pittsburgh, PA: U.S. Department of Labor, Mine Safety and Health Administration, Industrial Safety Division (ISD) report No. 06-213-90, August 1, 1990.

Stephan CR [1990b]. Omega 384 block as a seal construction material. Pittsburgh, PA: U.S. Department of Labor, Mine Safety and Health Administration, Industrial Safety Division (ISD) report No. 10-318-90, November 14, 1990.

Triebsch GF, Sapko MJ [1990]. Lake Lynn Laboratory: a state-of-the-art mining research laboratory. In: Proceedings of the International Symposium on Unique Underground Structures. Golden, CO: Colorado School of Mines, Vol. 2, pp. 75-1 to 75-21.

Weiss ES, Greninger NB, Perry JW, Stephan CR [1993a]. Strength and leakage evaluations for coal mine seals. In: Proceedings of the 25th International Conference on Safety in Mines Research Institutes. Pretoria, Republic of South Africa: Conference Papers for Day One, pp. 149-161.

Weiss ES, Greninger NB, Slivensky WA, Stephan CR [1993b]. Evaluation of alternative seal designs for coal mines. In: Proceedings of the Sixth U.S. Mine Ventilation Symposium, Salt Lake City, UT: University of Utah, chapter 97, pp. 635-640.

Weiss ES, Greninger NB, Stephan CR, Lipscomb JR [1993c]. Strength characteristics and air-leakage determinations for alternative mine seal designs. Pittsburgh, PA: U.S. Department of the Interior, Bureau of Mines, RI 9477.

Weiss ES, Slivensky WA, Schultz MJ, Stephan CR, Jackson KW [1996]. Evaluation of polymer construction material and water trap designs for underground coal mine seals. Pittsburgh, PA: U.S. Department of Energy, RI 9634.

Weiss ES, Slivensky WA, Schultz MJ, Stephan CR [1997]. Evaluation of water trap designs and alternative mine seal construction materials. In: Dhar BB, Bhowmick BC, eds. Proceedings of the 27th International Conference on Safety in Mines Research Institutes. Vol. 2. New Delhi, India: Oxford & IBH Publishing Co. Pvt. Ltd., pp. 973-981.

Weiss ES, Cashdollar KL, Mutton IVS, Kohli DR, Slivensky WA [1999]. Evaluation of reinforced cementitious seals. Pittsburgh, PA: U.S. Department of Health and Human Services, Public Health Service, Centers for Disease Control and Prevention, National Institute for Occupational Safety and Health, DHHS (NIOSH) Publication No. 99-136, RI 9647.

APPENDIX A.—SUMMARY TABLES OF AIR LEAKAGE MEASUREMENTS

Table A-1.—Air leakage measurements before the first explosion test (No. 354) of the HeiTech program

Location	Air leakage rates, m³/min, at pressure differential of—			
	0.17 kPa	0.32 kPa	0.50 kPa	0.86 kPa
Seal in crosscut 2	0.0	0.9	1.2	1.8
Seal in crosscut 3	1.1	1.7	2.2	2.9
Seal in crosscut 4	0.0	0.8	1.1	1.8
Seal in crosscut 5	0.0	0.8	1.0	1.7

Table A-2.—Air leakage measurements after the first explosion test (No. 354) of the HeiTech program

Location	Air leakage rates, m³/min, at pressure differential of—			
	0.16 kPa	0.30 kPa	0.51 kPa	1.01 kPa
Seal in crosscut 2	1.2	1.8	2.7	3.5
Seal in crosscut 3	17.0	21.9	31.6	37.5
Seal in crosscut 4	0.8	1.3	2.1	2.9
Seal in crosscut 5	0.9	1.3	1.9	2.7

Table A-3.—Air leakage measurements after sealant was reapplied and before the second explosion test (No. 355) of the HeiTech program

Location	Air leakage rates, m³/min, at pressure differential of—			
	0.21 kPa	0.35 kPa	0.55 kPa	1.06 kPa
Seal in crosscut 2	1.5	2.1	2.8	3.6
Seal in crosscut 3	0.0	<0.7	1.0	1.8
Seal in crosscut 4	0.9	1.3	1.7	2.7
Seal in crosscut 5	1.0	1.4	1.9	2.8

Table A-4.—Air leakage measurements after the second explosion test (No. 355) of the HeiTech program

Location	Air leakage rates, m³/min, at pressure differential of—			
	0.20 kPa	0.34 kPa	0.55 kPa	1.07 kPa
Seal in crosscut 2	1.6	2.1	2.8	3.8
Seal in crosscut 3	0.9	1.2	1.6	2.8
Seal in crosscut 4	1.1	1.4	2.2	3.2
Seal in crosscut 5	0.9	1.3	1.9	3.0

Table A-5.—Air leakage measurements before the first explosion test (No. 358) of the Barclay Mowlem program

Location	Air leakage rates, m³/min, at pressure differential of—			
	0.20 kPa	0.34 kPa	0.55 kPa	1.02 kPa
Seal in crosscut 2	0.0	0.0	0.0	1.1
Water stopping in crosscut 3 ...	13.3	19.1	27.4	—
	[1]4.8	[1]6.5	[1]9.3	[1]16.1

[1]Second air leakage test after re-securing water tubes that were dislodged during first leakage test.

Table A-6.—Air leakage measurements between the first (No. 358) and second (No. 359) explosion tests of the Barclay Mowlem program

Location	Air leakage rates, m³/min, at pressure differential of—			
	0.22 kPa	0.36 kPa	0.56 kPa	1.10 kPa
Seal in crosscut 2	1.1	1.5	2.1	3.0
Seal in crosscut 3	1.6	2.3	2.8	4.2

Table A-7.—Air leakage measurements between the second (No. 359) and third (No. 360) explosion tests of the Barclay Mowlem program

Location	Air leakage rates, m³/min, at pressure differential of—			
	0.10 kPa	0.26 kPa	0.47 kPa	1.06 kPa
Seal in crosscut 2	2.7	4.8	7.4	12.7
Seal in crosscut 3	2.4	3.6	5.2	8.4

Table A-8.—Air leakage measurements between the third (No. 360) and fourth (No. 361) explosion tests of the Barclay Mowlem program

Location	Air leakage rates, m³/min, at pressure differential of—			
	0.21 kPa	0.36 kPa	0.55 kPa	1.03 kPa
Seal in crosscut 2	3.0	4.3	5.8	8.6
Seal in crosscut 3	([1])	([1])	([1])	([1])
Overcast in B-drift at intersection with crosscut 3	1.0	1.0-1.6	1.6-1.9	2.4-2.8
Seal in crosscut 4	0.0	0.0	0.0	0.0

[1]Seal was destroyed by pressure pulse.

Table A-9.—Air leakage measurements between the fourth (No. 361) and fifth (No. 362) explosion tests of the Barclay Mowlem program

Location	Air leakage rates, m³/min, at pressure differential of—			
	0.21 kPa	0.36 kPa	0.55 kPa	1.03 kPa
Seal in crosscut 2	—	—	—	—
Overcast in B-drift at intersection with crosscut 3	1.7-2.2	2.5-2.8	2.8-3.7	4.5-5.3
	[1]1.6-2.1	[1]2.3-2.8	[1]3.2	[1]4.2-5.1
Seal in crosscut 4	0.0	0.0	<0.7	0.9
				[1]0.0

A dash (—) indicates that no data were measured.
[1]Second leakage test following gunite patching of overcast and crosscut 4 seal.

Table A-10.—Air leakage measurements before the seventh explosion test (No. 364) of the Barclay Mowlem program

Location	Air leakage rates, m³/min, at pressure differential of—			
	0.21 kPa	0.34 kPa	0.55 kPa	1.06 kPa
Seal in crosscut 2[1]	0.0	0.0	1.0	1.6
Seal in crosscut 3	0.0	0.0	1.0	1.7
Overcast in B-drift at intersection with crosscut 3	—	—	—	—
Seal in crosscut 4	0.0	0.0	0.0	0.9

A dash (—) indicates that no data were measured.
[1]Seal in crosscut 2 was re-gunited before air leakage test.

Table A-11.—Air leakage measurements after the seventh explosion test (No. 364) of the Barclay Mowlem program

Location	Air leakage rates, m³/min, at pressure differential of—			
	0.19 kPa	0.34 kPa	0.51 kPa	1.03 kPa
Seal in crosscut 2	1.0	1.4	2.0	2.9
Seal in crosscut 3	([1])	([1])	([1])	([1])
Overcast in B-drift at intersection with crosscut 3	—	—	—	—
Seal in crosscut 4	([1])	([1])	([1])	([1])

A dash (—) indicates that no data were measured.
[1]Seal was destroyed by pressure pulse.

Table A-12.—Air leakage measurements before the first explosion test (No. 365) of the Packsetter seal program with the solid concrete block

Location	Air leakage rates, m³/min, at pressure differential of—			
	0.21 kPa	0.36 kPa	0.52 kPa	1.02 kPa
Seal in crosscut 2	4.5	6.4	8.6	14.5
Seal in crosscut 3	6.7	9.3	12.3	18.9
Seal in crosscut 4	10.8	14.9	19.1	27.8

Table A-13.—Second air leakage measurements before the first explosion test (No. 365) of the Packsetter seal program with the solid concrete block

Location	Air leakage rates, m³/min, at pressure differential of—			
	0.21 kPa	0.37 kPa	0.57 kPa	1.05 kPa
Seal in crosscut 2	0.8	1.1	1.5	2.6
Seal in crosscut 3	0.9	1.3	1.8	3.0
Seal in crosscut 4	3.7	4.9	6.6	9.6
	—	[1]3.9	[1]5.0	[1]7.4

A dash (—) indicates that no data were measured.
[1]Air leakage rates obtained after reapplying sealant for a third time.

Table A-14.—Air leakage measurements between the first (No. 365) and second (No. 366) explosion tests of the Packsetter seal program with the solid concrete block

Location	Air leakage rates, m³/min, at pressure differential of—			
	0.22 kPa	0.36 kPa	0.55 kPa	1.02 kPa
Seal in crosscut 2	0.8	1.2	1.7	2.8
Seal in crosscut 3	1.3	1.8	2.6	3.5
Seal in crosscut 4	14.6	19.8	24.3	32.8

Table A-15.—Air leakage measurements after the second explosion test (No. 366) of the Packsetter seal program with the solid concrete block

Location	Air leakage rates, m³/min, at pressure differential of—			
	0.24 kPa	0.37 kPa	0.56 kPa	1.00 kPa
Seal in crosscut 2	0.9	1.3	1.8	2.8
Seal in crosscut 3[1]	1.7	2.4	2.9	4.5
Seal in crosscut 4	([2])	([2])	([2])	([2])

[1]Prior to this air leakage test, a small (unauthorized) amount of additional sealant was inadvertently applied by the vendor to seal 3 at the perimeter areas where the original sealant was dislodged and along the block/floor interface where the sealant had separated during the explosion.
[2]Seal was destroyed by pressure pulse.

APPENDIX B.—SUMMARY TABLES OF STATIC PRESSURE DATA FOR LLEM EXPLOSION TESTS

Table B-1.—HeiTech pumpable cementitious seals evaluation in the Lake Lynn Experimental Mine: pressure data, test 354 (November 6, 1997)

	TRANSDUCER				
Distance, ft (m)	Time of P_{max}, s	P_{max} 10-ms (15-pt) avg		Pressure-time integral ∫Pdt	
		psi	kPa	psi-s	kPa
13 (4.0)	0.510	39.0	270	20.0	138
59 (18.0)	0.495	28.5	200	6.8	47
84 (25.6)	0.500	28.0	193	7.0	49
134 (40.8)	0.520	27.0	185	6.4	44
184 (56.1)	0.550	29.0	200	6.8	47
234 (71.3)	0.578	24.0	165	5.2	36
304 (92.7)	0.614	23.0	160	4.8	33
403 (122.8)	0.678	22.0	150	3.9	27
501 (152.7)	0.744	17.0	120	4.2	29
	SEAL/STOPPING				
Location and distance, ft (m)	Type	P_{max} 10-ms (15-pt) avg		Pressure-time integral ∫Pdt	
		psi	kPa	psi-s	kPa
Seal in crosscut 2: 156 (47.5)	Static	28.0	190	6.6	46
Seal in crosscut 3: 246 (75.0)	Static	24.0	165	5.1	36
Seal in crosscut 4: 355 (108.2)	Static	22.5	155	4.3	30
Seal in crosscut 5: 452 (137.8)	Static	19.5	135	4.0	28

NOTE.—Pressure results are listed to nearest 0.5 psi and to nearest 5 kPa.
Pressure-time integral is calculated up to the time that the pressure trace returns to ~0 psi; it does not include the second (reflected) pressure pulse.
∫Pdt data are to nearest 0.1 psi-s and to nearest 1 kPa-s.

Table B-2.—HeiTech pumpable cementitious seals evaluation in the Lake Lynn Experimental Mine: pressure data, test 355 (November 20, 1997)

	TRANSDUCER				
Distance, ft (m)	Time of P_{max}, s	P_{max} 10-ms (15-pt) avg		Pressure-time integral ∫Pdt	
		psi	kPa	psi-s	kPa
13 (4.0)	0.717	39.0	270	22.2	153
59 (18.0)	0.677	24.0	165	6.7	46
84 (25.6)	0.685	25.0	175	7.3	50
134 (40.8)	0.688	27.0	185	6.3	43
184 (56.1)	0.714	28.0	195	5.7	39
234 (71.3)	0.735	24.5	170	5.3	36
304 (92.7)	0.780	23.5	160	4.9	34
403 (122.8)	0.841	20.0	140	3.7	26
501 (152.7)	0.910	16.0	110	1.8	12
	SEAL/STOPPING				
Location and distance, ft (m)	Type	P_{max} 10-ms (15-pt) avg		Pressure-time integral ∫Pdt	
		psi	kPa	psi-s	kPa
Seal in crosscut 2: 156 (47.5)	Static	27.5	190	6.0	41
Seal in crosscut 3: 246 (75.0)	Static	24.0	165	5.2	36
Seal in crosscut 4: 355 (108.2)	Static	22.0	150	4.3	30
Seal in crosscut 5: 452 (137.8)	Static	18.0	125	2.8	19

NOTE.—Pressure results are listed to nearest 0.5 psi and to nearest 5 kPa.
Pressure-time integral is calculated up to the time that the pressure trace returns to ~0 psi; it does not include the second (reflected) pressure pulse.
∫Pdt data are to nearest 0.1 psi-s and to nearest 1 kPa-s.

Table B-3.—Barclay Mowlem seal and stoppings evaluation in the Lake Lynn Experimental Mine: pressure data, test 358 (February 11, 1998)

Distance, ft (m)	TRANSDUCER					
	Time of P_{max}, s	P_{max} 10-ms (15-pt) avg		Pressure-time integral $\int Pdt$		
		psi	kPa	psi-s	kPa	
13 (4.0)	1.40	12.6	87	10.5	72.5	
59 (18.0)	1.31	4.2	29	1.5	10.5	
84 (25.6)	1.32	4.2	29	1.5	10.5	
134 (40.8)	1.36	4.2	29	1.3	9.0	
184 (56.1)	1.39	3.7	26	1.0	7.0	
234 (71.3)	1.41	2.9	20	0.8	5.5	
304 (92.7)	1.45	2.5	17	Small	Small	
403 (122.8)	1.54	1.8	13	Small	Small	
501 (152.7)	1.63	1.4	10	Small	Small	
598 (182.3)	1.89	1.3	9	Small	Small	
757 (230.7)	2.00	1.5	10	Small	Small	

Location and distance, ft (m)	SEAL/STOPPING					
	Type	P_{max} 10-ms (15-pt) avg		Pressure-time integral $\int Pdt$		
		psi	kPa	psi-s	kPa	
Seal in crosscut 2: 156 (47.5)	Static	4.0	27	1.3	9.0	
	Total	4.5	31	1.3	9.0	
Water stopping in crosscut 3: 246 (75.0)	Static	2.8	19	0.7	5.0	
	Total	3.3	23	0.9	6.0	
Quickseal stopping in crosscut 4: 355 (108.2)	Static	2.0	14	Small	Small	
	Total	—	—	—	—	

NOTE.—Pressures are listed to nearest 0.1 psi and to nearest 1 kPa.
Pressure-time integral is calculated up to the time that the pressure trace returns to ~0 psi; it does not include the second (reflected) pressure pulse.
$\int Pdt$ data are to nearest 0.1 psi-s and to nearest 0.5 kPa-s. "Small" refers to impulse <0.5 psi-s.
A dash (—) indicates that data were not available.

Table B-4.—Barclay Mowlem seals evaluation in the Lake Lynn Experimental Mine: pressure data, test 359 (February 27, 1998)

Distance, ft (m)	TRANSDUCER					
	Time of P_{max}, s	P_{max} 10-ms (15-pt) avg		Pressure-time integral $\int Pdt$		
		psi	kPa	psi-s	kPa	
13 (4.0)	0.58	36.0	250	23.5	162	
59 (18.0)	0.55	27.0	185	6.1	42	
84 (25.6)	0.52	29.0	200	6.4	44	
134 (40.8)	0.54	31.5	215	5.7	39	
184 (56.1)	0.58	28.0	190	3.6	25	
234 (71.3)	0.59	26.0	180	4.2	29	
304 (92.7)	0.65	18.5	125	2.7	19	
403 (122.8)	0.74	15.0	100	2.0	14	
501 (152.7)	0.80	11.5	80	0.9	6	
598 (182.3)	0.86	8.0	55	0.7	5	
757 (230.7)	0.98	5.5	40	—	—	

Location and distance, ft (m)	SEAL/STOPPING					
	Type	P_{max} 10-ms (15-pt) avg		Pressure-time integral $\int Pdt$		
		psi	kPa	psi-s	kPa	
Seal in crosscut 2: 156 (47.5)	Static	30.0	205	4.8	33	
	Total	39.5	270	5.4	37	
Seal in crosscut 3: 246 (75.0)	Static	25.0	170	3.9	27	
	Total	32.0	220	3.8	26	

NOTE.—Pressures are listed to nearest 0.5 psi and to nearest 5 kPa.
Pressure-time integral is calculated up to the time that the pressure trace returns to ~0 psi; it does not include the second (reflected) pressure pulse.
$\int Pdt$ data are to nearest 0.1 psi-s and to nearest 1 kPa-s.
A dash (—) indicates that data were not available.

Table B-5.—Barclay Mowlem seals evaluation in the Lake Lynn Experimental Mine: pressure data, test 360 (March 3, 1998)

Distance, ft (m)	TRANSDUCER				
	Time of P_{max}, s	P_{max} 10-ms (15-pt) avg		Pressure-time integral $\int Pdt$	
		psi	kPa	psi-s	kPa
13 (4.0)	0.78	90.0	620	55.0	379
59 (18.0)	0.76	57.5	395	15.0	103
84 (25.6)	0.72	67.0	460	16.0	110
134 (40.8)	0.70	59.5	410	14.7	101
184 (56.1)	0.70	47.0	320	11.0	79
234 (71.3)	0.68	67.5	465	19.0	131
304 (92.7)	0.67	73.0	505	8.0	55
403 (122.8)	0.71	62.0	430	9.0	62
501 (152.7)	0.75	41.5	285	7.5	52
598 (182.3)	0.80	24.5	170	5.8	40
757 (230.7)	0.89	14.0	100	5.8	40

Location and distance, ft (m)	SEAL/STOPPING				
	Type	P_{max} 10-ms (15-pt) avg		Pressure-time integral $\int Pdt$	
		psi	kPa	psi-s	kPa
Seal in crosscut 2: 156 (47.5)	Static	54	370	13.1	90
	Total	55	380	14.8	102
Seal in crosscut 3: 246 (75.0)	Static	69	475	—	—
	Total	79	545	[1]~8	[1]~55

[1]Integral up to time of failure.

NOTE.—Pressures are listed to nearest 0.5 psi and to nearest 5 kPa.
Pressure-time integral is calculated up to the time that the pressure trace returns to ~0 psi; it does not include the second (reflected) pressure pulse.
$\int Pdt$ data are to nearest 0.1 psi-s and to nearest 1 kPa-s.
A dash (—) indicates that data were not available.

Table B-6.—Barclay Mowlem seals and overcast evaluation in the Lake Lynn Experimental Mine: pressure data, test 361 (March 26, 1998)

Distance, ft (m)	TRANSDUCER				
	Time of P_{max}, s	P_{max} 10-ms (15-pt) avg		Pressure-time integral $\int Pdt$	
		psi	kPa	psi-s	kPa
13 (4.0)	1.49	4.0	27	1.5	10.5
59 (18.0)	1.46	3.5	24	1.2	8.5
84 (25.6)	1.48	3.3	23	1.1	7.5
134 (40.8)	1.50	3.2	22	1.0	7.0
184 (56.1)	1.56	3.0	21	0.9	6.0
234 (71.3)	1.59	2.3	16	0.7	5.0
246 (75.0)	1.62	2.4	16	0.8	5.0
304 (92.7)	1.65	2.3	16	0.7	5.0
403 (122.8)	1.72	1.9	13	0.5	3.5
501 (152.7)	1.81	1.3	9	0.5	3.5
598 (182.3)	1.87	1.0	7	0.5	3.5
757 (230.7)	2.16	1.1	8	0.5	3.5

Location and distance, ft (m)	SEAL/STOPPING				
	Type	P_{max} 10-ms (15-pt) avg		Pressure-time integral $\int Pdt$	
		psi	kPa	psi-s	kPa
Seal in crosscut 2: 156 (47.5)	Static	3.1	21	1.0	6.5
	Total	3.5	24	1.0	7.0
Overcast, crosscut 3 and B-drift intersection: 246 (75.0)	Static	2.3	16	0.6	4.0
Seal in crosscut 4: 355 (108.2)	Static	2.1	14	0.6	4.0
	Total	2.3	16	0.6	4.0

NOTE.—Pressures are listed to nearest 0.1 psi and to nearest 1 kPa.
Pressure-time integral is calculated up to the time that the pressure trace returns to ~0 psi; it does not include the second (reflected) pressure pulse.
$\int Pdt$ data are to nearest 0.1 psi-s and to nearest 0.5 kPa-s.

Table B-7.—Barclay Mowlem seals and overcast evaluation in the Lake Lynn Experimental Mine: pressure data, test 362 (March 31, 1998)

	TRANSDUCER				
Distance, ft (m)	Time of P_{max}, s	P_{max} 10-ms (15-pt) avg		Pressure-time integral $\int Pdt$	
		psi	kPa	psi-s	kPa
13 (4.0)	1.18	9.0	62	8.2	56.5
59 (18.0)	0.91	6.5	44	1.9	13.0
84 (25.6)	0.96	6.5	43	1.8	12.5
134 (40.8)	0.98	6.0	42	1.6	10.5
184 (56.1)	1.02	5.7	39	1.2	8.5
234 (71.3)	1.04	4.1	28	1.0	7.0
246 (75.0)	1.08	4.2	29	1.1	7.5
304 (92.7)	1.09	4.2	29	1.1	7.5
403 (122.8)	1.16	3.8	26	0.7	4.5
501 (152.7)	1.24	2.7	19	0.6	4.0
598 (182.3)	1.32	1.8	13	0.6	4.0
757 (230.7)	1.62	1.8	13	0.6	4.0
	SEAL/STOPPING				
Location and distance, ft (m)	Type	P_{max} 10-ms (15-pt) avg		Pressure-time integral $\int Pdt$	
		psi	kPa	psi-s	kPa
Seal in crosscut 2: 156 (47.5)	Static	5.9	41	1.4	9.5
	Total	6.8	47	1.5	10.5
Overcast, crosscut 3 and B-drift intersection: 246 (75.0)	Static	4.3	30	0.8	6.0
Seal in crosscut 4: 355 (108.2)	Static	4.0	28	0.9	6.0
	Total	4.7	33	0.9	6.5

NOTE.—Pressures are listed to nearest 0.1 psi and to nearest 1 kPa.
Pressure-time integral is calculated up to the time that the pressure trace returns to ~0 psi; it does not include the second (reflected) pressure pulse.
$\int Pdt$ data are to nearest 0.1 psi-s and to nearest 0.5 kPa-s.

Table B-8.—Barclay Mowlem seals and overcast evaluation in the Lake Lynn Experimental Mine: pressure data, test 363 (April 1, 1998)

	TRANSDUCER				
Distance, ft (m)	Time of P_{max}, s	P_{max} 10-ms (15-pt) avg		Pressure-time integral $\int Pdt$	
		psi	kPa	psi-s	kPa
13 (4.0)	0.88	16.5	114	12.0	82.0
59 (18.0)	0.83	9.5	66	2.3	15.5
84 (25.6)	0.82	9.4	65	2.2	15.0
134 (40.8)	0.84	8.9	61	1.9	12.5
184 (56.1)	0.88	8.9	61	1.5	10.0
234 (71.3)	0.88	6.8	47	1.1	7.5
246 (75.0)	0.92	5.8	40	1.2	8.5
304 (92.7)	0.95	6.3	44	1.3	8.5
403 (122.8)	1.02	5.9	41	0.8	5.5
501 (152.7)	1.07	4.3	30	0.7	5.0
	SEAL/STOPPING				
Location and distance, ft (m)	Type	P_{max} 10-ms (15-pt) avg		Pressure-time integral $\int Pdt$	
		psi	kPa	psi-s	kPa
Seal in crosscut 2: 156 (47.5)	Static	8.9	61	1.7	11.5
	Total	11.1	77	1.8	12.5
Overcast, crosscut 3 and B-drift intersection: 246 (75.0)	Static	6.8	47	0.9	6.0
Seal in crosscut 4: 355 (108.2)	Static	6.1	42	1.0	7.0
	Total	7.1	49	1.1	7.5

NOTE.—Pressures are listed to nearest 0.1 psi and to nearest 1 kPa.
Pressure-time integral is calculated up to the time that the pressure trace returns to ~0 psi; it does not include the second (reflected) pressure pulse.
$\int Pdt$ data are to nearest 0.1 psi-s and to nearest 0.5 kPa-s.

Table B-9.—Barclay Mowlem seals and overcast evaluation in the Lake Lynn Experimental Mine: pressure data, test 364 (April 3, 1998)

Distance, ft (m)	Time of P_{max}, s	TRANSDUCER P_{max} 10-ms (15-pt) avg		Pressure-time integral $\int Pdt$	
		psi	kPa	psi-s	kPa
13 (4.0)	0.57	38.0	265	21.0	145
59 (18.0)	0.56	27.0	185	5.4	37
84 (25.6)	0.57	27.0	190	5.6	39
134 (40.8)	0.59	29.0	200	5.1	35
184 (56.1)	0.62	27.0	185	5.0	34
234 (71.3)	0.64	23.0	160	3.7	26
246 (75.0)	0.72	24.0	165	3.4	23
304 (92.7)	0.67	17.0	115	3.0	21
403 (122.8)	0.77	17.0	115	1.8	12
501 (152.7)	0.84	11.5	80	1.1	8

Location and distance, ft (m)	Type	SEAL/STOPPING P_{max} 10-ms (15-pt) avg		Pressure-time integral $\int Pdt$	
		psi	kPa	psi-s	kPa
Seal in crosscut 2: 156 (47.5)	Static	28.0	195	5.0	34
	Total	35.0	240	4.8	33
Seal in crosscut 3: 246 (75.0)	Static[1]	23.0	160	—	—
	Total	29.0	200	[2]1.4	[2]10
Overcast, crosscut 3 and B-drift intersection: 246 (75.0)	Static	6.0	41	1.1	8
Seal in crosscut 4: 355 (108.2)	Static	17.0	115	—	—
	Total	20.0	135	[2]1.3	[2]9

[1]Weighted average of 234 ft and 304 ft, then averaged with 246 ft.
[2]Integral up to time of failure.

NOTE.—Pressures are listed to nearest 0.5 psi and to nearest 5 kPa.
Pressure-time integral is calculated up to the time that the pressure trace returns to ~0 psi; it does not include the second (reflected) pressure pulse.
$\int Pdt$ data are to nearest 0.1 psi-s and to nearest 1 kPa-s.
A dash (—) indicates that data were not available.

Table B-10.—Packsetter solid-concrete-block seals evaluation in the Lake Lynn Experimental Mine: pressure data, test 365 (June 22, 1998)

Distance, ft (m)	Time of P_{max}, s	TRANSDUCER P_{max} 10-ms (15-pt) avg		Pressure-time integral $\int Pdt$	
		psi	kPa	psi-s	kPa
13 (4.0)	0.548	34.0	235	23.0	159
59 (18.0)	0.517	23.5	165	4.9	34
84 (25.6)	0.521	22.5	155	5.0	34
134 (40.8)	0.547	22.0	150	4.5	31
184 (56.1)	0.571	22.0	155	4.0	28
234 (71.3)	0.619	18.5	130	3.2	22
304 (92.7)	0.650	19.5	135	2.8	19
403 (122.8)	0.713	16.0	110	1.7	12
501 (152.7)	0.771	11.5	80	0.9	6

Location and distance, ft (m)	Type	SEAL/STOPPING P_{max} 10-ms (15-pt) avg		Pressure-time integral $\int Pdt$	
		psi	kPa	psi-s	kPa
Seal in crosscut 2: 156 (47.5)	Static	22.0	150	4.3	30
	Total	27.5	190	4.6	32
Seal in crosscut 3: 246 (75.0)	Static	18.5	130	3.1	21
	Total	~30	~210	3.3	23
Seal in crosscut 4: 355 (108.2)	Static	17.5	120	2.2	15
	Total	~25	~170	2.3	16

NOTE.—Pressure results are listed to nearest 0.5 psi and to nearest 5 kPa.
Pressure-time integral is calculated up to the time that the pressure trace returns to ~0 psi; it does not include the second (reflected) pressure pulse.
$\int Pdt$ data are to nearest 0.1 psi-s and to nearest 1 kPa-s.

Table B-11.—Packsetter solid-concrete-block seals evaluation in the Lake Lynn Experimental Mine: pressure data, test 366 (June 25, 1998)

\multicolumn{6}{c}{TRANSDUCER}					
Distance, ft (m)	Time of P_{max}, s	P_{max} 10-ms (15-pt) avg		Pressure-time integral ∫Pdt	
		psi	kPa	psi-s	kPa
13 (4.0)	0.562	37.0	255	—	—
59 (18.0)	0.590	24.0	165	6.1	42
84 (25.6)	0.584	24.5	170	6.4	44
134 (40.8)	0.585	25.5	175	5.4	37
184 (56.1)	0.610	27.5	190	5.1	35
234 (71.3)	0.640	22.5	160	4.1	28
304 (92.7)	0.680	20.5	140	3.4	23
403 (122.8)	0.737	16.5	115	1.9	13
501 (152.7)	0.800	12.0	85	1.1	8
\multicolumn{6}{c}{SEAL/STOPPING}					
Location and distance, ft (m)	Type	P_{max} 10-ms (15-pt) avg		Pressure-time integral ∫Pdt	
		psi	kPa	psi-s	kPa
Seal in crosscut 2: 156 (47.5)	Static	26.5	185	5.3	36
	Total	35.5	245	5.6	39
Seal in crosscut 3: 246 (75.0)	Static	22.5	155	4.0	27
	Total	~32	~220	4.1	28
Seal in crosscut 4: 355 (108.2)	Static	18.5	125	2.6	18
	Total	~27	~190	2.6	18

NOTE.—Pressure results are listed to nearest 0.5 psi and to nearest 5 kPa.

Pressure-time integral is calculated up to the time that the pressure trace returns to ~0 psi; it does not include the second (reflected) pressure pulse.

∫Pdt data are to nearest 0.1 psi-s and to nearest 1 kPa-s.

A dash (—) indicates that data were not available.

APPENDIX C.—SUMMARY TABLE OF FLAME ARRIVAL DATA FOR LLEM EXPLOSION TESTS

Table C-1.—HeiTech and Packsetter seals evaluation in the Lake Lynn Experimental Mine: flame arrival time data

Flame sensor distance, ft (m)	Flame arrival time, s			
	Test 354	Test 355	Test 365	Test 366
13 (4.0)	NA	0.34	0.20	0.30
84 (25.6)	NA	0.64	0.48	0.52
134 (40.8)	NA	0.66	0.52	0.56
184 (56.1)	NA	0.72	ND	0.60
234 (71.3)	NA	0.88	ND	0.76
304 (92.7)	NA	Small	ND	1.10
403 (122.8)	NA	ND	ND	ND
598 (182.3)	NA	ND	ND	ND

NA Data not available.
ND No detectable signal.
"Small" means that the signal was <1 V.

NOTE.—Flame arrival time corresponds to ≥1-V signal on flame sensor. Data are relative to ignition time.

Table C-2.—Barclay Mowlem seals, stoppings, and overcast evaluation in the Lake Lynn Experimental Mine: flame arrival time data

Flame sensor distance, ft (m)	Flame arrival time, s						
	Test 358	Test 359	Test 360	Test 361	Test 362	Test 363	Test 364
13 (4.0)	1.120	0.279	0.291	0.940	0.535	0.432	0.315
84 (25.6)	1.287	0.503	0.505	1.482	0.941	0.793	0.547
134 (40.8)	ND	0.541	0.549	ND	Small	0.888	0.584
184 (56.1)	ND	0.608	0.595	ND	ND	Small	0.637
234 (71.3)	ND	0.798	0.631	ND	ND	ND	0.875
304 (92.7)	ND	0.940	0.680	ND	ND	ND	Small
403 (122.8)	ND	ND	0.825	ND	ND	ND	ND
598 (182.3)	ND	ND	ND	ND	ND	ND	ND

ND No detectable signal.
"Small" means that the signal was <1 V.

NOTE.—Flame arrival time corresponds to ≥1-V signal on flame sensor. Data are relative to ignition time.

APPENDIX D.—SUMMARY TABLES OF LVDT DISPLACEMENT DATA FOR LLEM EXPLOSION TESTS

Table D-1.—Barclay Mowlem seals evaluation in the
Lake Lynn Experimental Mine:
LVDT data, test 358
(February 11, 1998)

Location and instrument	Maximum displacement, mm
Seal in crosscut 2:	
LVDT Upper	0.2
LVDT Middle	0.3
LVDT Bottom	0.1
LVDT Right	0.2

Table D-2.—Barclay Mowlem seals evaluation in the
Lake Lynn Experimental Mine:
LVDT data, test 359
(February 27, 1998)

Location and instrument	Maximum displacement, mm
Seal in crosscut 2:	
LVDT Upper	2.3
LVDT Middle	5.6
LVDT Bottom	1.8
LVDT Right	10.7
Seal in crosscut 3:	
LVDT Upper	0.8
LVDT Middle	2.0
LVDT Bottom	1.3
LVDT Right	4.5

Table D-3.—Barclay Mowlem seals evaluation in the
Lake Lynn Experimental Mine:
LVDT data, test 360
(March 3, 1998)

Location and instrument	Maximum displacement, mm
Seal in crosscut 2:	
LVDT Upper	4.0
LVDT Middle	5.6
LVDT Bottom	3.1
LVDT Right	12.2
Seal in crosscut 3:[1]	
LVDT Upper	17.9
LVDT Middle	32.5
LVDT Bottom	31.1
LVDT Right	29.3

[1]Destroyed.

Table D-4.—Overcast LVDT data, test 361 (March 26, 1998)

Instrument	Maximum displacement, mm	
Deck:	Up	Down
Middle	5.5	0.9
Toward C-drift	3.0	1.4
Outby	1.2	0.2
Side wall, outby:	Outby	Inby
Top	≤0.1	≤0.1
Middle	≤0.1	≤0.1
Middle-Right	≤0.1	≤0.1
Bottom	≤0.1	≤0.1
Wing wall:	Toward A-drift	Toward C-drift
Middle	~0.2	~0.2

Table D-5.—Overcast LVDT data, test 362 (March 31, 1998)

Instrument	Maximum displacement, mm	
Deck:	Up	Down
Middle	14.5	15.7
Toward C-drift	10.8	13.7
Outby	12.9	14.8
Side wall, outby:	Outby	Inby
Top	0.6	0.0
Middle	0.5	0.0
Middle-Right	0.4	0.0
Bottom	0.5	0.0
Wing wall:	Toward A-drift	Toward C-drift
Middle	0.0	1.4

Table D-6.—Overcast LVDT data, test 363 (April 1, 1998)

Instrument	Maximum displacement, mm	
Deck:	Up	Down
Middle	15.1	15.7
Toward C-drift	16.4	14.6
Outby	15.2	15.6
Side wall, outby:	Outby	Inby
Top	0.7	—
Middle	0.7	0.2
Middle-Right	—	—
Bottom	0.4	0.1
Wing wall:	Toward A-drift	Toward C-drift
Middle	0.6	2.6

A dash (—) indicates that data were not available.

Table D-7.—Seal LVDT data, test 363 (April 1, 1998)

Location and instrument	Maximum displacement, mm
Seal in crosscut 4:	
LVDT Middle	~0.1

Table D-8.—Overcast LVDT data, test 364 (April 3, 1998)

Instrument	Maximum displacement, mm	
	Up	Down
Deck:		
Middle	15.8	14.0
Toward C-drift	7.4	1.4

Table D-9.—Seal LVDT data, test 364 (April 3, 1998)

Location and instrument	Maximum displacement, mm
Seal in crosscut 3:	
LVDT Middle	12.6
Seal in crosscut 4:	
LVDT Middle	17.0

www.ingramcontent.com/pod-product-compliance
Lightning Source LLC
Chambersburg PA
CBHW081903170526
45167CB00007B/3133